Concrete

Building Pathology

Also of interest

Building Pathology
Principles and Practice
David S. Watt
0-632-04875-1

Building Maintenance Management
Barrie Chanter & Peter Swallow
0-632-05766-1

Lee's Building Maintenance Management
Fourth Edition
Paul Wordsworth
0-632-05362-3

Dampness in Buildings
Alan Oliver
Second Edition revised by
J. Douglas & J.S. Stirling
0-632-04085-8

Concrete

Building Pathology

Edited by
Susan Macdonald

Foreword by
David Watt & Peter Swallow

Blackwell
Science

© 2003 by Blackwell Science Ltd,
a Blackwell Publishing Company
Editorial Offices:
Osney Mead, Oxford OX2 0EL, UK
  Tel: +44 (0)1865 206206
Blackwell Science, Inc., 350 Main Street,
Malden, MA 02148-5018, USA
  Tel: +1 781 388 8250
Iowa State Press, a Blackwell Publishing
Company, 2121 State Avenue, Ames, Iowa
50014-8300, USA
  Tel: +1 515 292 0140
Blackwell Publishing Asia Pty Ltd,
550 Swanston Street, Carlton South, Victoria
3053, Australia
  Tel: +61 (0)3 9347 0300
Blackwell Wissenschafts Verlag,
Kurfürstendamm 57, 10707 Berlin, Germany
  Tel: +49 (0)30 32 79 060

The right of the Authors to be identified as
the Authors of this Work has been aserted in
accordance with the Copyright, Designs and
Patents Act 1988.

First published 2003 by Blackwell Science
Ltd

Library of Congress
Cataloging-in-Publication Data
is available

ISBN 0-632-05251-1

A catalogue record for this title is available
from the British Library

Set in 10/13pt Palatino
by SNP Best-set Typesetter Ltd., Hong Kong
Printed and bound in Great Britain by
TJ International, Padstow, Cornwall

For further information on
Blackwell Science, visit our website:
www.blackwell-science.com

Contents

Contributors

Bill Addis MA (Cantab), PhD, MCIOB, Comp. Member IStructE
While a university academic for over 15 years, Bill Addis lectured on building engineering design and construction, and the history of building materials, structures and building engineering to both engineers and architects. He has written widely on the history of construction and building, especially the history of structural engineering design. In 1997 he helped devise and curate a major exhibition at the Pompidou Centre on The Art of the Engineer celebrating some 200 years of construction and engineering history. He now works as a consulting engineer with Buro Happold, advising engineers, architects and clients on how best to reduce the environmental impact of construction.

Ranjit Bassi BSc, MSc, PhD, DIC
Ranjit Bassi joined the Building Research Establishment in 1990 after completing his PhD at Imperial College. He has given seminars on the selection of coatings, concrete repair and the implications of European Standards to the British Cement Association training courses, the Oil Colour and Chemistry Association, the International Concrete Repair Association and BRE external seminars. He was presented with the Owen Nutt award by the Federation of Epoxy Resin Formulators and Applicators for the best paper on concrete research in 1996.

John Boxall BSc (Hons), FTSC, MICorr, MIMF
Following a period in industry, John Boxall joined the Building Research Establishment in 1973, eventually becoming a Projects Manager responsible for contracts on surface coatings for construction substrates. He is currently a Director of the Paint Advisory Bureau, an independent technical consultancy serving the needs of clients in the coating industry and its customer base.

John Broomfield DPhil, EurIng, FICorr, FIM
John Broomfield is an independent corrosion consultant specialising in the corrosion of steel in concrete. He has published over 60 papers on this

subject, as well as a patent and a book published in 1997. He runs an international practice and designed some of the earliest cathodic protection systems on bridges and buildings in the UK, Hong Kong and Australia. His latest work includes assessing a cathodic protection design for an historic steel-framed building in Scotland, corrosion monitoring systems on bridges and concrete pavements in England and Singapore, and trials of innovative designs of cathodic protection systems on wharves and motorway bridges.

Michael Bussell BSc (Eng), MASCE

Michael Bussell worked as a structural engineer with Ove Arup & Partners for 30 years, and is now an independent consultant. His experience includes both the design of numerous concrete structures and, more recently, their investigation and repair when distress or deterioration has occurred. He has a long-standing interest in the history of building structures and in their appraisal for adaptive re-use. He contributed two papers on the use and development of reinforced concrete construction in Britain to a dedicated issue of the Proceedings of the Institution of Civil Engineers on concrete history (since published in book form). He is also the author of a guide to the appraisal of existing iron and steel structures, which was published in 1997 by the Steel Construction Institute.

Paul E. Gaudette BSCE, FACI

Since joining Wiss, Janney, Elstner Associates, USA, 18 years ago, Paul Gaudette has been involved in a wide range of investigations and repair projects. Most of his experience has been in investigative work and the design of repairs for various types of modern and historic concrete structures. His recent work includes the repair of Mies van der Rohe's Promontory Apartments building in Chicago, John Earley's Baha'i House of Worship in Wilmette, Illinois, the Franklin Delano Roosevelt Memorial in Washington, DC, and the Centre Street Bridge (Calgary, Alberta, Canada). He is a Fellow of the American Concrete Institute and is instructor for a number of the ACI courses. He is also an active member of the Association for Preservation Technology International (APT).

Lance Horlyck BE (Hons), Grad Dip Management, MIEAust, CPeng

Lance Horlyck is a structural engineer with over 20 years experience in the design and assessment of a broad range of different structures. For the past 10 years he has worked as a private consultant providing specialist assistance to Sydney Water, Australia. His work on the rehabilitation of sewer and water infrastructures has included a number of concrete structures from the late 19th century, where old concretes and heritage issues have played a major role in the repair strategy.

Harry J. Hunderman MArch, FAIA, FAPT

Harry Hunderman is a conservation architect and a principal and Architecture Group manager with Wiss, Janney, Elstner Associates, USA, and is past president of the Association for Preservation Technology (APT). Harry Hunderman has specialised in the application of technology to the conservation of the built environment, and has special expertise in providing problem-solving, rehabilitation and repair services for new and historic buildings. His recent work includes the repair of the Promontory Apartments building in Chicago, the preservation of Cape Hatteras Lighthouse in North Carolina, various projects for the Kennedy Center in Washington, DC, and the stabilisation and repair of the Pullman State Historic Site in Chicago.

John Lambert BE (Hons), MEngSc (Struct), MIE Aust, CPEng, Member AISC

John Lambert has over 30 years of experience in the design of reinforced concrete, steel and prestressed concrete structures for civil, industrial and commercial developments. Before becoming Senior Structural Engineer at Sydney Water, Australia, he was employed by consulting engineering firms on a wide range of structures, including 5 years in Canada on high-rise developments. He has been responsible for the rehabilitation of many existing water and sewerage facilities for Sydney Water, particularly important heritage structures, including large reservoirs and aqueducts.

Susan Macdonald BSc (Arch), BArch, MA (Conservation Studies), RIBA

Susan Macdonald is an architect who recently returned to her native Australia, where she is the Assistant Director of the NSW Heritage Office. Before this she spent 4 years working within the Architectural Conservation Team of English Heritage, which is responsible for advice, research and publications concerned with practical building conservation. She has a particular interest in the conservation of twentieth-century buildings and materials, and has published a number of articles on this subject. She is the Editor of *Modern Matters: Principles and Practice in Conserving Recent Architecture* and *Preserving Post-War Heritage: the Care and Conservation of Mid-Twentieth Century Architecture*. Susan is a former Secretary of DOCOMOMO UK, and a former member of the executive committee of ICOMOS Australia.

Bryan Marsh BSc, PhD, CEng, MICE, FICT, FCS

Bryan Marsh is an Associate with Arup Research & Development. He is a chartered civil engineer and concrete technologist with over 20 years experience. Prior to joining Arup in 1997, he was Head of Concrete Technology at BRE. He has published over 30 papers on concrete technology.

He is a member of the British Standards Institution Committee on Concrete and an Honorary lecturer in the School of Civil Engineering at the Queen's University, Belfast. He is a member of the Council and Chairman of the Technical & Education Committee of the Institute of Concrete Technology.

Stuart Matthews BEng, PhD, CEng, MICE, MIStructE, MCIWEM
Stuart Matthews is Director of the Centre for Concrete Construction, Construction Division of BRE. He is a chartered civil and structural engineer with extensive experience in the fields of civil engineering and building, structural assessment, non-destructive testing and investigation. Whilst at the BRE, Stuart Matthews has undertaken research into the in-service behaviour and appraisal of existing structures, methods of investigation and structural assessment, and full-scale structural testing and monitoring, as well as concrete rehabilitation and maintenance methods. He has produced a number of publications on these topics. He is also a member of several national and international technical committees (Fib, RILEM, ISO etc.).

John R. Morlidge BSc (Hons), MSc, PhD, MICorr
John Morlidge is a Senior Corrosion Consultant within the Centre for Concrete Construction at BRE. He has undertaken various corrosion research projects focusing on the maintenance, rehabilitation and repair of reinforced concrete structures. These have included investigations into the use of concrete corrosion inhibitors, electrochemical monitoring of steel-reinforced concrete structures and the improved use of concrete patch repair materials. He is also a member of several national and international corrosion technical committees.

Matthew Murray BSc, MSc
Matthew Murray is currently a research manager at ECOTEC working on the evaluation of urban regeneration schemes. Previously, Matthew worked at BRE for several years on project managing and undertaking built-environment research programmes. His projects included developing and publishing best-practice checklists and guidance for specifiers of building conservation and cleaning projects, assessing the impact of pollution on buildings, and the environmental impact and economics of laser cleaning historic buildings in Scotland. As part of his responsibilities at BRE, Matthew was seconded to the DETR's Research Analysis and Evaluation team engaged in a critical evaluation of component life-cycle models of social housing stock repair investment.

Tony Sheehan BA (Oxon), MIM, CEng

Tony Sheehan is an Associate Director of Ove Arup and Partners, and provides advice to internal and external clients on the use of materials in building and civil engineering. Advice is provided at all stages of a project, from initial design to trouble-shooting during construction (including the critical appraisal of new products), and investigating failures and defects in completed buildings and structures. He has particular expertise in non-metallic materials (notably concrete and polymeric materials) and smart materials, and he uses his expertise to re-direct theoretical and academic developments towards applications in the construction industry. Tony is a member of the Concrete Society Council, and is on various Concrete Society Committees.

Deborah Slaton MArch, MA

Deborah Slaton is an architectural conservator with Wiss, Janney, Elstner Associates, USA. She specialises in the investigation, repair and conservation of older and modern historic structures. Representative projects for which she has served as architectural conservator include Cape Hatteras Lighthouse in North Carolina, the Buffalo Conservatory in Buffalo, New York, and the Pennsylvania State Memorial at Gettysburg National Military Park. She is currently vice president of the Historic Preservation Education Foundation. She is co-author of a book on specifying for conservation projects, which is published jointly by CSI and the Association for Preservation Technology International.

Foreword

Concrete has a history stretching back as far as 7000 BC, but following the collapse of the Roman Empire its use declined and consequently its value as a versatile structural meterial was largely forgotten. Interest in concrete did not reawaken until the eighteenth century, when engineers began to confront the construction of lighthouses and other marine structures on difficult sites and with challenging environments. The story of modern concrete effectively begins, however, with Joseph Aspdin's patent for the manufacture of Portland cement in 1824 and notion, as early as the 1830s, of reinforcing Portland cement concrete with iron rods and bars. During the latter half of the nineteenth century, practical techniques for reinforcing concrete began to emerge and the structural possibilities for this new material explored, culminating in the erection of Weaver's flour mill at Swansea in 1898, the first multi-storey concrete-framed building in Britain.

During the twentieth century, growth in the use of structural concrete was rapid as engineers and architects began to realise its potential and produced innovative and groundbreaking designs for both civil engineering and building structures alike. Unfortunately, the significance of many of these structures, both in terms of their engineering technology and architectural design, was not recognised and significant numbers have been lost to demolition, affected by unsympathetic alteration, or allowed to deteriorate through lack of maintenance. Only in the last 20 years or so have the most significant structures been afforded statutory protection and considered worthy of conservation.

The problem facing conservators today is just how to deal with reinforced concrete and its architectural and engineering heritage. Whilst there have been developments in treatments to halt the damaging effects of corroding reinforcement, which are appropriate for the normal building stock, they potentially compromise the aesthetic or design authenticity of landmark buildings and structures.

It is hoped that this book will disseminate to as wide an audience as possible, providing a sound understanding of the technical issues concerning concrete deterioration as well as a timely review of the repair techniques available. In so doing, it may – in some small way – help to inform the on-going debate over the philosophical difficulties raised by the repair and conservation of concrete.

David S. Watt
Peter Swallow
University of De Montfort

Preface

This book is one of a number of books on building pathology dealing with specific diagnosis, prognosis and repair issues associated with concrete buildings and structures. Building pathology embraces a holistic approach to the repair of buildings and structures. This involves a detailed understanding of how the structure is built, the materials of which it is constructed, how it has been used, how it has performed over time, and all the factors that have affected its current condition.

The use of concrete in one form or another dates back thousands of years, but it is only in the last century that reinforced concrete has come into widespread use. Since the industrialisation of cement production in the nineteenth century, concrete has become one of the most widely used construction materials in the world. The problems associated with the deterioration of concrete are widely known, and well-developed codes of practice now define its use. This book provides a basis for understanding buildings and structures constructed in concrete by tracing its historical development as a preliminary to recognising problems and selecting appropriate methods of repair and remediation. It is based on the assumption that in order to identify repair options correctly, a detailed understanding is necessary of the original component materials and construction methods, and of the environmental and use factors that have affected the building since its construction. A glossary is provided at the end of the book as a ready reference to the terms used throughout.

The case studies were selected to illustrate the various repair methods that are available for the repair of concrete, and all exemplify the building pathology approach. They have been selected from around the world, and most deal with buildings or structures that are protected by heritage legislation. The repair strategies adopted reflect the special care needed when dealing with heritage buildings, rendering more conventional care unacceptable.

The book is a collaborative effort, drawing together the collective wisdom of experts in the history, defect diagnosis and remediation of concrete. The editor would like to thank all the contributors for their assistance in the development of the book and for their individual contributions.

The content of Chapter 6 draws upon valuable contributions and comments made by colleagues within the Building Research Establishment (BRE) and some other organisations, and upon the findings of a number of previous research programmes. Funding for much of BRE's research work was provided by the Department for the Environment, Transport and the Regions and the Highways Agency.

Thanks are due to David Watt, the author of *Building Pathology: Principles and Practice*, who first had the idea for this book and who has played such an important role in the dissemination of information within the various heritage conservation professions.

I am particularly grateful to those who provided case studies, mostly at quite short notice, including Lance Horlyck and John Lambert for their contribution entitled *The repair of two nineteenth-century concrete structures in Sydney, Australia*. Thanks are also due to Mount Ida Press, and to Harry Hunderman, Deborah Slaton and Paul Gaudette for enabling us to reprint their article on the Mies van der Rohe Promontory Apartments in Chicago, that first appeared in the APT Journal's special edition *Mending the Modern*. The case study on the National War Memorial Carillon Tower in Wellington, New Zealand, is based on a paper presented at the Australasian Corrosion Association Annual Meeting, Auckland, New Zealand (2000), by John Broomfield, W. R. Jacob of Global Corrosion Ltd (UK), and W. L. Mandeno and S. M. Bruce of Opus International Consultants Ltd (NZ).

I am grateful for the experience gained whilst working within the Architectural Conservation Team at English Heritage. Here I was able to see a number of newly identified listed buildings, constructed over the last 100 years, that were suffering from various defects and forms of damage and decay. The difficulties of achieving suitable repair strategies that were appropriate for listed buildings was in many cases extremely difficult and aroused my interest in this subject. Figure 5.1 is developed from work done by Kevin Davies for English Heritage in 1998. Both John Broomfield and I acknowledge Kevin's encouragement and assistance.

The patience and assistance of Julia Burden, our publisher, is gratefully acknowledged. I would like to thank my family for their on-going forbearance and the sacrifices they made to enable me to finish the book.

Unless otherwise stated, the photographs are by the author of the chapter in which they appear.

Chapter 1

Introduction

Susan Macdonald

Despite being thought of as a modern material, concrete has been in use for thousands of years. Lime-based hydraulic cements (those set by chemical reaction with water) have been in use since Roman times, and examples of ancient mortars (cement and aggregate) still survive today. The word *concrete* comes from the Latin *concretus*, which means mixed together or compounded (Stanley, 1979).

The use of concrete has changed dramatically over the last 100 years as our understanding about the material and how it is made has improved. The industry continues to evolve today as experience grows and its versatility and cost-effectiveness is exploited. The design, construction method, materials used and standards of workmanship employed in the creation of a concrete structure will vary enormously according to the date of its construction. All these factors will affect the durability of the structure, and may result in particular deterioration problems that correlate to the construction period. It is thus imperative to understand the characteristics of a concrete structure (including its history) and the factors that have affected it since its construction, in order to inform the investigation of problems, reach a correct diagnosis and specify appropriate repair techniques (Fig. 1.1).

Concrete: composition and characteristics

Concrete is made up of aggregates of various sizes, broadly categorised as fine (commonly sand) and coarse (typically crushed stone or gravel), combined with a cement paste (a mixture of cement and water) which acts as the binder. The greater proportion of concrete is aggregate, known as the filler, which is bulky and relatively cheaper than the cement. As the constituents of concrete derive from stone, it has been known variously as artificial stone, cast stone, reconstructed stone and reconstituted stone. However, concrete must be thought of as a material distinct from stone, with its own very different durability and weathering characteristics and repair requirements.

1

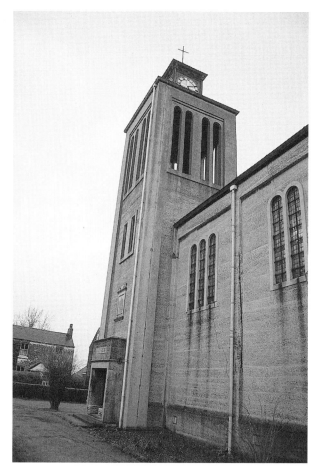

Fig. 1.1 St John and St Mary Magdalene, Goldthorpe, South Yorkshire, 1914–1916, designed by A. Nutt. The problems associated with the reinforced concrete decay in this church are a result of the particular way it was constructed, and are different from problems that have occurred in the building shown in Fig. 1.3.

Binders

Portland cement, in its various guises (rapid hardening, sulfate-resistant, low heat and so on), is the binder most commonly used in modern concrete. Portland cement is commercially manufactured by blending limestone or chalk with clays that contain alumina, silica, lime, iron oxide and magnesia. These compounds are heated together to high temperatures (1300–1500°C). During this process, the calcium carbonate in the limestone ($CaCO_3$) loses carbon dioxide (CO_2), leaving the lime (CaO). The products in the clays, such as alumina and silica, are what provides the hydraulic properties. The heating process chemically combines these materials through partial melting of the silicates and vitrification on cooling to form compounds of silicates, aluminates and aluminosilicates in the form of a clinker. This is then finely ground to a powder.

Most Portland cement clinkers contain four principal compounds (CPA, 1998);

- tricalcium silicate ($3CaOSiO_2$),
- dicalcium silicate ($2CaOSiO_2$),
- tricalcium aluminate ($3CaOAl_2O_3$),
- a series containing iron ($2CaOFe_2O_3$–$6CaO.2Al_2O_3.Fe_2O_3$).

Other cementitious binders used as partial replacements for Portland cement include (ground granulated) blast-furnace slag (produced as a by-product of smelting iron ore), pulverised fly ash (a by-product of coal-fired power generation) and silica fume cement replacement materials. The use of different binder types varies the characteristics of the concrete, in particular its permeability to water and hence its durability, as discussed later. Replacement binders, small amounts of additives or changes in the chemical composition of the cement create concretes for specific purposes such as rapid hardening/high early strength, sulfate resistance and low heat generation.

Hydration

Concrete is made by mixing cement paste with aggregate. Concrete's strength is derived from the hydration process, i.e. the reaction between the binder (the cementitious component) and water, which leads to the formation of an alkali paste that surrounds and binds the aggregate together as a solid mass.

During the hydration process, the calcium silicates and aluminates in the cement cement react with water to form crystals of hydrated lime and a gel of of hydrated calcium silicates and calcium aluminates in the following reactions (Torraca, 1988):

$$3CaO\ SiO_2 \quad\quad + H_2O \rightarrow Ca(OH)_2 \text{ and } 3CaO\ SiO_2$$

Tricalcium silicate + water $\rightarrow$ hydrated lime and

tricalcium silicate

(with less calcium than the initial compound)

$2CaO SiO_2$ $+ H_2O$ $\rightarrow$ $Ca(OH)_2$ and $2CaO SiO_2$

Dicalcium silicate + water $\rightarrow$ hydrated lime and
tricalcium silicate

(with less calcium than the initial compound)

$3CaO Al_2O_3$ $+ H_2O$ $\rightarrow$ $Ca(OH)_2$ and $3CaO Al_2O_3$

Tricalcium aluminate + water $\rightarrow$ hydrated lime and
tricalcium aluminate

(with less calcium than the initial compound)

The second part of this process occurs when the water penetrates the gel coating the solid particles and the hydration process continues inside, resulting in fibrous shoots that form a network and causing the material to solidify (Torraca, 1988). The lime formed in this process slowly converts to calcium carbonate. Small amounts of gypsum are added to the cement as part of the manufacturing process to slow the process of hydration.

The ratio of water to the binder, the binder contents, curing and compaction are the most important factors controlling the quality of concrete, although there are others such as the use of sound, durable and well-graded aggregates. The permeability of concrete is an important factor in its durability. As explained in Chapter 4 in more detail, permeable concrete is more susceptible to the environmental effects that reduce durability and result in deterioration.

Curing

The amount of water used in a concrete mix governs its resultant permeability and strength. Excess water beyond that required in the hydration process does not become chemically combined and evaporates, resulting in pores that provide channels for air and water, allowing atmospheric gases (particularly oxygen, water vapour and carbon dioxide) to penetrate more easily. Hence, excess water results in more permeable concrete. Too little water results in stiff mixes that are difficult to work and place, and so chemical admixtures are used in concrete manufacture to reduce the water content whilst maintaining the workability of the mix.

To achieve low permeability and satisfactory strength, freshly made concrete needs to be kept moist, and also shaded from sun and wind, to prevent it from drying out too rapidly. This process is known as curing. Curing allows the hydration process to continue for a longer period, which ensures that the hydrated cement fills the pores that were originally filled with water. This reduces the voids within the concrete that allow the passage of air and moisture and exacerbate deterioration (see Chapter 4). Rapid water loss during the hydration process can reduce the plasticity of the concrete and result in cracking. Typical curing methods include retaining formwork in place, covering with damp hessian or plastic sheeting, and applying an impermeable membrane to the exposed concrete surface, usually by spraying.

Compaction

Compaction of the wet concrete mix drives out trapped air and brings surplus water to the surface where it can evaporate. This will result in greater strength and reduced permeability. In the past, this was carried out by hand using ramming or 'punning' tools, but the more thorough compaction obtainable by machinery led to the development of vibrating 'pokers' placed into the concrete, while external vibrators clamped to the formwork are used in making precast concrete.

Reinforced concrete

Concrete is very strong in compression but relatively weak in tension. To overcome this deficiency when concrete is used as a structural building material, and to combat early shrinkage and control subsequent diurnal thermal expansion and contraction, reinforcement (historically iron and later steel) is included in areas where tension occurs to create reinforced concrete. Steel and concrete have similar coefficients of thermal expansion and form an effective composite section. Concrete is mixed wet, poured around the placed reinforcement into the formwork (which forms the required concrete profile and also supports its wet weight), and compacted or vibrated to expel air. It then sets or hardens by the chemical reaction explained above.

Reinforced concrete is a manufactured material, produced either in a casting yard (in the case of precast concrete) or, more commonly for buildings, it is site-made (when it is known as in situ concrete). Its performance and appearance are dependent on the individual materials of which it is composed, the standard of workmanship during construction, and subsequent environmental conditions and maintenance. Improved quality control and mix designs, along with a now recognised understanding of some of the earlier degradation problems, have, in recent decades, greatly reduced the chances of poor quality construction and premature decay.

Unfortunately, a number of concrete buildings and structures exhibit problems that are often related to the practice methods (workmanship,

Fig. 1.2 St John and St Mary Magdalene, Goldthorpe, South Yorkshire. Despite the reasonably thick walls, the location of the reinforcement is variable and does not appear to have been well restrained horizontally, as can be seen in the photograph. The concrete is quite porous, and the building is suffering from prolonged deterioration with scant maintenance, which has resulted in some structural problems. Understanding the level of knowledge and types of materials used at the time when it was built is important to an analysis of the problems and may influence the repair decisions.

understanding about best practice) and materials (type of reinforcement, cement matrix, water content) used at the time of their construction. Chapter 2 provides an outline of the development of concrete, including the different materials and methods used at different times, and explains the reasons why concretes from various periods deteriorate in the way they do (Fig. 1.2). The repair methods used for concrete have also developed rapidly over the last 100 years, and particularly in the last 30 years.

Concrete repair is now a major, fully fledged, multinational industry. There are many organisations devoted to the research and development of concrete repair materials and methods, and some of these are listed in Appendix A.

Building pathology

Building pathology is a term used to describe the holistic approach to understanding buildings (Watt, 1999). It is based on the principle that to repair and maintain a building effectively, a detailed understanding is required of how it was designed, constructed, used and changed, and how the particular environmental, material and structural conditions have affected it. Therefore, this book begins with an outline of the historic development of concrete as a building material, progresses to how to assess a concrete building, explains how concrete decays, describes common defects and damage, provides information on the repair methods available, and discusses maintenance issues. The final chapter discusses concrete in the future. Case studies have been used to highlight this approach for a range of different problems and solutions.

General information about the building pathology approach is available in the first volume of this series, which deals with building performance, survey and assessment, and discusses the general principles of remediation and repair, building management and aftercare (Watt, 1999).

Historic buildings and structures

Over the last 10–15 years there has been a growing interest in conserving buildings from the twentieth century, many of which made extensive use of reinforced concrete. Consequently, this has generated discussion about the conservation of statutorily protected concrete buildings, centering on the difficulty of utilising typical repair methods whilst also observing the usual conservation principles of conserve as found, minimal intervention and reversibility. The key issues relate to the conservation aims of minimal intervention and reducing the visual impact of repair on the existing building. At present, reinforced concrete repair tends to be an invasive process owing to the way the material decays and the resultant damage. It is difficult to achieve a technically proficient repair without changing the appearance of the building. This is discussed in more detail in Chapter 5, and case studies are provided in Chapter 8 which illustrate how the building pathology approach may assist in dealing with some of these issues.

The framework provided by the legal statutes and policies that protect historic buildings are perfectly aligned with the concept of building pathology. The most important principle is that in order to develop appropriate conservation strategies, a thorough understanding of the building's heritage significance, causes of deterioration and future maintenance and care commitments are required. There are specific controls for listed buildings and buildings within conservation areas in the UK. Consent is required to alter a listed building if the proposed works will affect its character as a building or structure of historical interest. Concrete repair is not yet specifically included within the suites of guidance documents provided by the key heritage agencies such as English Heritage and Historic Scotland. However, there are now a number of case studies that have informed their approach and demonstrated the relevance of their stated conservation principles to concrete conservation (Fig. 1.3).

In addition to the statutory Acts and policies that protect historic buildings and structures, there are a number of key documents that guide

Fig. 1.3 The Loggia, Bungalow 'A', Whipsnade, Buckinghamshire, 1933–1936, designed by Berthold Lubetkin. The thin-walled construction of the architect's own house has suffered problems typical of its era and construction type. The slender precast-concrete members of the loggia and fenestration have sagged. Despite the thin-walled construction, with its minimal cover to the reinforcing steel, the concrete is of a high quality and has therefore survived relatively well. The building was to be repaired using a fairly minimal approach.

conservation practice in the UK and further afield that will influence the conservation of concrete buildings and structures. These include the ones listed below.

- *The Venice Charter* (1966): the International Council on Monuments and Sites (ICOMOS) inaugural international charter introducing the principles for conserving monuments and sites.
- *The Burra Charter* (1981 and amended 1999): ICOMOS Australia's response to the Venice Charter that guides conservation practice in that country.
- *British Standard 7913* (BSI, 1998). *Guide to the Principles of the Conservation of Historic Buildings*: describes the principles of conservation.
- *The Repair of Historic Buildings: Advice on Principles and Methods* (Brereton, 1995): English Heritage's guiding principles for conservation practice.
- *SPAB (Society for the Protection of Ancient Buildings) Manifesto* (1877): the original manifesto written by William Morris that underpins the organisation's work.

In addition to the statutory and principle documents mentioned above, organisations such as DOCOMOMO International (the international working party for the documentation and conservation of buildings, sites and neighbourhoods of the Modern Movement), ICOMOS (International Council of Monuments and Sites) and APT (Association for Preservation Technology International) have been instrumental in forwarding research and discussion on the conservation of concrete buildings. There are now a number of publications (mainly conference and seminar proceedings) that deal with this topic, and these are included in the list of further reading at the end of this chapter. Appendix A includes contact details for the various organisations mentioned above.

Sustainable development

Local Agenda 21 is the blueprint for local communities to set out actions that contribute to global sustainability. It presumes that local activities have a major effect on the environment, and so uses local government as the base for the development of local strategies that promote environmental, social and economic sustainability.

Given the emphasis placed on the reduction of construction waste and the importance of the re-use of existing resources, it seems probable that the culture of demolition of more recent buildings and their frequent replacement may require rethinking. Aggregate production for concrete uses a finite resource, and there are particular environmental issues associated with aggregate quarrying that are likely to become more problematic in the future. In the past, it has been argued that the particular deterioration characteristics of concrete sometimes make it unviable to remediate a concrete building, and that demolition is inevitable. Chapter 6 discusses the life-cycle assessment of concrete buildings in detail. It is only by applying the principles of whole-life costing that a true assessment of a building's viability in the future can be made. Chapter 7 discusses sustainable development issues in more detail.

References

Brereton, C. (1995) *The Repair of Historic Buildings: Advice on Principles and Methods.* English Heritage, London.

BSI (1998) *Guide to the Principles of the Conservation of Historic Buildings.* BS7913, British Standards Institution, London.

CPA (1998) *Reinforced Concrete: History, Properties and Durability.* CPA Monograph No. 1, Corrosion Prevention Association, Aldershot.

ICOMOS (1966) *International Charter for the Conservation and Restoration of Monuments and Sites (The Venice Charter).* International Council on Monuments and Sites, Venice.

ICOMOS Australia (1999) *Charter for Places of Cultural Significance (The Burra Charter).* International Council on Monuments and Sites, Canberra.

SPAB (Society for the Protection of Ancient Buildings) (1877) *Manifesto.* SPAB, London.

Stanley, C. (1979) *Highlights in the History of Concrete.* British Cement Association, Crowthorne.

Torraca, G. (1988) *Porous Building Materials.* International Centre for the Study of the Preservation and the Restoration of Cultural Property (ICCROM), Rome.

Watt, D. (1999) *Building Pathology: Principles and Practice.* Blackwell Science, Oxford.

Further reading

Allen, J. (1994) The conservation of modern buildings. In: Mills, E. (ed) *Building Maintenance and Preservation: A Guide to Design and Management.* Revised edition, Butterworth–Heineman, London.

Bronson, S. and Jester, T. (eds) (1997) *Special Issue: Mending the Modern.* APT Bulletin **28**:4.

Burman, P., Garner, K. and Schmidt, L. (eds) (1996) *The Conservation of 20th Century Historic Buildings.* Institute of Advanced Architectural Studies, York.

De Jonge, W. and Doolaar, A. (1997) *The Fair Face of Concrete: Conservation and Repair of Exposed Concrete.* Preservation Dossier No. 2, DOCOMOMO International, Eindhoven.

Department of the Environment (1994) *Planning Policy Guidance Note 15: Historic Buildings and Conservation Areas*. HMSO, London.

DOCOMOMO International Conference Proceedings: First international conference, September 1990, Eindhoven; Second international conference, September 1992, Dessau, Germany; Third international conference, September 1994, Barcelona; Fourth international conference, September 1996, Bratislava, Slovakia; Fifth international conference, September 1998, Stockholm. DOCOMOMO, Delft.

English Heritage (1997) *Sustaining the Historic Environment: New Perspectives on the Future*. English Heritage, London.

Everett, A. (1986) *Mitchell's Building Series: Materials*. Mitchell Publishing, London.

Grattan, D.W. (ed) (1993) *Saving the Twentieth Century: The Conservation of Modern Materials*. Proceedings of the 1991 Conference, Canadian Conservation Institution, Ottawa.

Historic Scotland (1998) *Memorandum of Guidance on Listed Buildings and Conservation Areas*. Historic Scotland, Edinburgh.

HMSO (1994) *Sustainable Development: The UK Strategy*. HMSO, London.

International Council on Monuments and Sites (ICOMOS) Seminar on 20th Century Heritage, Helsinki, 18–19 June 1995, Working papers. (Recommendations and Council of Europe Principles developed from the conference are available through the ICOMOS web-site on http://www.icomos.org)

Macdonald, S. (ed) (1996) *Modern Matters: Principles and Practice in Conserving Recent Architecture*. Donhead, Shaftesbury.

Macdonald, S. (ed) (2001) *Preserving Post-War Heritage: The Care and Conservation of Mid-Twentieth Century Architecture*. Donhead, Shaftesbury.

Morton, W. Brown III, Hume, G.I., Weeks, K.D. and Jandl, H.W. (1992) *The Secretary of the Interior's Standards for Rehabilitation and Illustrated Guidelines for Rehabilitating Historic Buildings*. US Department of the Interior *et al.*, Washington, DC.

Slaton, D. and Foulks, G. (2000) *Preserving the Recent Past 2*. Historic Preservation Education Foundation *et al.*, Washington, DC.

Slaton, D. and Shiffer, R.A. (ed) (1995) *Preserving the Recent Past*. Historic Preservation Education Foundation, Washington, DC.

Stratton, M. (ed) (1997) *Structure and Style: Conserving Twentieth Century Buildings*. E. and F.N. Spon, London.

Part One

The Nature and Use of Concrete Buildings and Structures

Chapter 2

Key Developments in the History of Concrete Construction and the Implications for Remediation and Repair

Bill Addis & Michael Bussell

Introduction

The term 'concrete' denotes many things. There are records of concrete being used 5000 years ago, as well as in Roman times, in Mediaeval Europe and in eighteenth and nineteenth century Britain and France, and since about 1900 concrete has become a principal construction material in every country of the world.

However, all these versions of concrete were different. They had their own unique mixtures of ingredients and an enormous variety of durability and strength characteristics. Much Roman concrete was very similar to modern concrete in that it could set under water. Most concrete or mortar from the Middle Ages was very weak, and used lime in place of the later Portland cement. It often served as little better than a muddy filler to ensure that compressive loads were better conveyed between uneven blocks of stone. In the mid-eighteenth century, Smeaton rediscovered Roman hydraulic cements that can fully cure under water. This development was seen as a vital military secret and the object of much French espionage in order that they too could build better harbours to serve their navy.

Nowadays, when people refer to a concrete bridge or building, they are usually using the word as shorthand for 'reinforced concrete', i.e. concrete reinforced by steel bars carefully placed in order to carry the tensile and shearing stresses in structural elements. The properties of reinforced concrete are even more complex than those of concrete, since they depend not only on the properties of the concrete and steel, but also upon the effectiveness with which loads are conveyed between the steel bars and the concrete. This combination of the high-tensile strength of steel and the compressive strength of concrete, together with the ability to cast shapes of extraordinary variety and complexity, has resulted in a remarkable change in our architecture since it first came to be exploited in the last decade of the nineteenth century.

A more recent development was prestressed concrete, in which the steel is tensioned before loads are applied to the concrete. This allows the use of more slender sections, and also reduces or eliminates cracking of the concrete which, however fine it may be, inevitably occurs when reinforced concrete is subject to tension or bending.

During the twentieth century there was a steady rise in the strength of ordinary concrete. This came about as the chemical processes became better understood, along with the importance of ensuring the right conditions for the concrete to be made. Before the First World War, concrete strengths of $11-15\,N/mm^2$ were typical, although the mix designs were specified not in terms of strength, but by the relative volumes of cement, fine aggregate (sand) and coarse aggregate (stone or gravel), a practice that survives today among small builders. Typical mixes were 1:2:4, 1:1.5:3 and 1:1:2. In large structures it soon became necessary to specify the required strength of the concrete. Strength was usually measured by crushing a cube of the concrete, giving rise to the term 'cube strength' as the common criterion for strength. By the 1930s, typical cube strengths had risen to $15-20\,N/mm^2$, and since the 1950s they have risen again to $20-30\,N/mm^2$. Since the 1930s it has been possible, using special mixes, to make higher-grade concretes which nowadays reach strengths of up to $120\,N/mm^2$, i.e. more than four times that of ordinary grade concrete.

During the same period there was also a growing understanding of what influences the durability of concrete and reinforced concrete. Especially important is the need to protect the embedded reinforcement against corrosion arising from contact with atmospheric oxygen and moisture. The durability must match the exposure conditions, and is improved by increasing the thickness of the concrete cover around the reinforcement. The protection also improves if the cement content is increased and the water–cement ratio is decreased. These mixes also produce greater concrete strength, which nowadays can almost be a by-product of meeting the requirements for greater durability.

The key lesson from this overview is that whenever it is proposed to re-use a structure of any age that contains concrete, it is imperative to investigate the strength and other properties of the concrete and to establish whether, and precisely how and where, it has been reinforced with steel.

This chapter reports many of the key events in the development of concrete, but it does not address the reasons why they occurred. Many strands of economic and cultural history have affected the progress of concrete besides the obvious striving for better structural properties and lower costs. These complex and fascinating stories are well covered elsewhere (Bowley, 1966; Guillerme, 1986; Simmonet, 1992a; Powell, 1996). One observation from these histories is worth noting. Generally speaking, the building industry has been very conservative—clients, architects, engineers and contractors alike—in both initiating and adopting innovations. The diffusion of new ideas through the construction industry was, indeed still is, usually from bridge and civil engineering projects to industrial buildings, then to major commercial buildings, and finally to the 'architectural' end of the market.

The 'pre-history' of modern concrete: up to the 1890s

The use of concrete

Context

The eighteenth century brought a considerable need for construction in Britain. International trade required more, larger and better ports, not only for the cargo vessels, but also for the growing navy which was needed to defend the shipping routes and the colonies. The distribution of goods within Britain increased the demand for roads, navigable rivers and canals, warehouses and docks. By the end of the century, the industries that used the imported raw materials were producing enormous quantities of goods, not only for home consumption, but also for export. The increasing affluence that resulted from this growing activity led not only to private wealth and more opulent houses in town and country, but also to the beginning of public wealth and the opportunities for towns to compete with one another by demonstrating their grandeur in built form.

The nineteenth century brought more—much more—of the same. The most dramatic change was brought about by the development of the railways from the late 1820s. By the middle of the century, every town in Britain had been transformed—some would say destroyed—by at least one and often several railway stations and the lines that fed them. The railways soon came to serve not only to transport goods and people between major towns, but also for shorter journeys, and the age of the commuter was firmly established by the end of the century. This increased mobility also transformed the needs of workplaces (bespoke offices were constructed from the 1860s onwards) and opened up the commercial potential for places of mass entertainment, from the Crystal Palace and the Albert Hall down to small theatres, music halls and museums.

Architecture

Mainstream architecture was hardly touched by the use of concrete until after the end of the nineteenth century. For small buildings, masonry construction was ubiquitous, and when concrete was used it was disguised as masonry. While large buildings often had an internal frame employing cast iron or, from the 1860s, wrought iron columns and beams, they were required to have masonry, load-bearing external walls until the early twentieth century. However, these bare facts conceal the growing use of concrete during the century, especially from about 1870.

The first paper presented at the newly formed Royal Institute of British Architects, in 1836, was on the nature of concrete and its use in buildings up to that time (Godwin, 1836). This paper had perhaps been stimulated in 1835 by the first all-concrete house in Swanscombe, Kent. However, another paper in 1875 noted that there had been little development in the architectural treatment of concrete since the earlier paper (Payne, 1875). This author also noted that while the architectural debate had previously been about 'the apparatus for moulding walls, various kinds of building blocks, and compositions for making concrete', it had now changed to 'turning into artistic account' the apparatus for moulding and constructing concrete walls, roofs, arches, etc. and the proper way to ornament these 'in accordance with the peculiar properties and nature of the material'. The author also noted the 'special advantages' of using iron 'in connection with concrete'.

Another view from around the same time was less optimistic: 'in an architectural point of view it [concrete] was a failure entirely. . . . In the construction of walls, warehouses or other buildings where no architectural display was required, it would be an advantage. Anything in the way of architectural decoration which [the author had] seen in concrete buildings had been either cement mouldings or cornices and the appearance was poor and wretched.' Nevertheless, from the early 1870s, nearly every week in the technical press there were examples of houses made substantially from concrete (Fig. 2.1). Although they were few as a proportion of all houses, they indicate a growing interest in the new material. A factor in this interest was surely cost: concrete could be cheaper than traditional masonry construction.

Many architects had embraced developments in using unreinforced concrete via their builders. It had been used for walls in domestic buildings in the 1850s, and by 1870 this was quite common. However, by far the greatest use of concrete in buildings during the last third of the nineteenth century was in the floors of buildings. Although these usually offered no direct, visible contribution to the architecture, they did enable builders to achieve long spans, especially when used to span between, and to envelop, iron beams. This practice also provided fire resistance to the otherwise vulnerable iron beams.

Precast concrete was also developed in the 1870s for architectural use, especially by William Lascelles (Morris, 1978; Hurst, 1996). His system of

Fig. 2.1   A mass concrete house: Fenlands Villa, Chertsey, Surrey (1870) (from *The Builder*, 12 February 1870, p. 125).

panels, about $1\,m \times 0.75\,m \times 40\,mm$ thick, was used for many domestic and industrial buildings, and usually featured ornate decoration both for external use in walls and internally for ceilings (Fig. 2.2). As so often happened, the mouldability of concrete was used to disguise the material by giving it the appearance of stone or even brick tiles.

The role of architects in bringing about these radical changes to the way buildings were constructed was slight. They were dependent on what the builders and contractors might offer and, increasingly during the 1870s and 1880s, a growing number of patented systems that competed for the attention of architects and builders alike.

Structure

Throughout the century, most concrete was used in civil engineering projects, with which architects usually had little involvement, e.g. maritime, river and canal-side structures, industrial buildings, some bridges and, of course, many, many foundations (Chrimes, 1996b). From the 1820s many buildings, bridges and other structures built near water had concrete foundations. Telford's St Katharine Docks in London had a 300-mm bed of concrete beneath the dock wall, as did most similar structures built during the expansion of London's port in the 1850s and 1860s (Sharp,

Fig. 2.2   The Buffet at Royal Albert Dock using Lascelles system of precast concrete panels (*Building News*, 19th October 1883).

1996). The Royal Albert Dock (1876–1880) had mass concrete walls. From the 1870s, concrete also came to be used as a waterproof cut-off in dams as a replacement for the traditional puddled clay.

These structures generally use concrete to carry compression forces only and are used in situations where dead-weight is seldom crucial, and is often an advantage. For these reasons the concrete was seldom stressed very highly, and it can hardly be claimed that it was being used to the limit of its potential structural performance.

In buildings it was a different story. Concrete made its greatest impact in floor structures, where it was always appropriate to reduce the amount of concrete used to a minimum for structural as well as financial reasons. The brick jack-arch, so common in mills and other factory buildings from the 1790s, was often adapted or copied using concrete, for instance using brickwork with a concrete fill and topping, or mass concrete, or hollow clay blocks spanning between parallel iron beams. These contain wrought-iron tie-rods, sometimes exposed and sometimes embedded, to locate the beams during construction and to carry the arch thrusts should an adjacent arch be (or become) absent. A simpler version of this idea is found in filler joist construction, in which the full depth between two I-beams is filled with concrete and works as a flat arch (Crook, 1965; Stanley, 1979; Hurst, 1996). These various types of floor were generally marketed as fireproof flooring systems, and those employing iron were variously described as 'iron elements protected by concrete' or 'concrete floors strengthened by iron', depending on the way the inventor looked at it. Many patents were taken out for such 'fireproof floors' using concrete, although many of these found little use (Hurst, 1998).

Even more surprising than the early use of iron with concrete is the frequent use of unreinforced concrete. One warehouse with concrete walls reached a height of 60 feet (18.3 m), and floor spans of several metres were quite common right up to the end of the nineteenth century. Documented examples include a 175-mm flat slab spanning 8 m × 6 m, and an arch slab, varying from 275 mm down to 75 mm thick, spanning 15 m × 3.6 m and carrying 'an immense weight of machinery and men' in a warehouse. Another of similar thickness, over a drawing room, was reported in 1876 as being 40 feet by 16 feet (12.2 m × 4.9 m), and 4.5 inches (115 mm) thick at the crown and 280 mm thick at the springing. No less alarming was a balcony projecting 1.22 m from the wall supported on cantilevers, cast in situ, and tapering from 280 mm to 75 mm. The only means adopted for fixing was a cemented butt joint between the cantilevers and the house wall. Twelve days after casting the balcony was successfully tested by three men running along it. It is worth mentioning here because it is more than likely that some of these remain, yet to be discovered by unsuspecting architects and engineers.

Concrete construction and design

Materials

All that the Romans had discovered and developed in the art of making concrete was lost when the empire faded, and the story of concrete in modern Britain begins again, more or less from scratch, in the late Middle Ages. From this time concrete was used mainly in foundations, for instance at Salisbury Cathedral, and as a lime mortar in masonry construction. These cements were usually unable to set in the presence of water, and this generally limited their use to buildings.

Until the mid-eighteenth century, concrete and mortar used in traditional buildings was what we now call lime concrete and lime mortar. Lime is formed by burning chalk or limestone at about 900°C to produce quicklime. To make mortar, this quicklime is slaked with water to form hydrated lime, and sand is added to form a mortar of suitable consistency. The properties of such a mortar vary with the source of the lime and the precise nature of other materials added to the mix. Pure lime hardens by reaction with atmospheric carbon dioxide to form calcium carbonate. This is a slow process, and cannot occur under water. Hence, such lime is called 'non-hydraulic'.

From the sixteenth century, however, it was also well known that a few natural limes produced mortar that would set under water—the so-called hydraulic limes. These contained clay, and when burnt with the lime produced a cementitious material that underwent a partial set when mixed with water. A similar effect could be achieved by using the natural cementitious material *pozzolana*. This is a volcanic ash, originally employed by the Romans, which takes its name from the town of Pozzuoli near Naples. It was imported into England from Holland from the seventeenth century for use when hydraulic cement was needed. The great breakthrough for concrete came during the eighteenth century when the precise nature of this chemical process was studied and understood.

Having visited the Netherlands in the mid-1750s, John Smeaton began a series of experiments to find the best lime mixture to produce the hydraulic cement he needed for his Eddystone Lighthouse, off the coast of Plymouth. He settled on equal measures of limestone from Aberthaw in South Wales and imported pozzolana.

One particular source of limestone that produced a hydraulic cement was the foreshore of the Thames Estuary near the Isle of Sheppey, where so-called 'cement stones' or 'septaria' were found. These were discovered by James Parker, who patented the material in 1796, calling it 'Roman cement', and the word 'cement' was used to signify this material until the middle of the century, when it came to refer to its successor 'Portland cement'.

Portland cement was patented by Joseph Aspdin, a bricklayer from Leeds, in 1824, although it only became a truly reliable product in 1845 when another cement manufacturer successfully managed to control the high temperatures involved. It is made by heating the chalk and clay ingredients to a much higher temperature than in Parker's process. He called it Portland to liken it in people's minds to the famous stone from Portland, which it resembled in colour (Skempton, 1966).

The success of concrete depends crucially on the ratios of water and cement in the mix, but it also depends on the cleanliness of the ingredients—the water, which must be of drinking quality, the sand and other aggregates. To this day, concrete is notoriously vulnerable to careless manufacture. Throughout its history, there have been many structural failures due to poor manufacture, often by unscrupulous contractors who boost their profits by skimping on both cement and supervision of the process. The weekly press in the 1870s had regular reports of concrete failures, followed by letters from the owners of the patented systems arguing that it was not the patent or the material that was at fault, but the unskilled labour or ignorant contractors who were to blame.

Perhaps the first suggestion of an idea for reinforcing concrete was a statement by J.C. Loudon in the *Encyclopaedia of Cottage, Farm, and Village Architecture* in 1830, in which it was suggested that flat roofs might be constructed of a latticework of iron tie-rods thickly embedded in cement, and cased with flat tiles. However, the earliest patent for reinforcing concrete (with disused cables from coal mines) to make a beam or floor was in 1854 by W.B. Wilkinson, a plasterer and manufacturer of cement and 'artificial stone' products from Newcastle upon Tyne (Fisher Cassie, 1955;

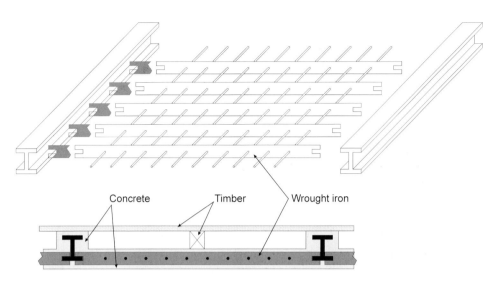

Fig. 2.3   Hyatt's patent reinforced concrete floor system developed about 1877 (drawing Bill Addis).

Brown, 1966). His firm placed adverts such as the one below in the trade press right up to the end of the century:

> W.B. Wilkinson Est. 1841. Concrete workers. Designers and constructors of concrete staircases and fireproof floors of all descriptions, with as little iron as possible, and that wholly in tension, thereby preventing waste.

During the following 30 years many similar patents were filed (de Courcy, 1987; Simmonet, 1992b). In 1877, Thaddeus Hyatt, an American working in Britain, privately published the results of extensive experiments he had undertaken on concrete beams reinforced with iron which had been tested at Kirkaldy's materials testing laboratory in London. A seven-storey building designed in 1886 using Hyatt's patented system of reinforcing concrete survives to this day at 63 Lincoln's Inn Fields in London (Figs. 2.3 and 2.4) (Hurst, 1996). The scene was already set for the development of modern reinforced concrete.

Construction methods

The essence of forming any concrete element is to pour the viscous mix into a mould—the formwork—that defines its external profile, and to leave this in place until the concrete has hardened sufficiently for the formwork to be safely removed. This may be done at the required location of the element, when the work is described as 'in situ' concrete. Alternatively, the element may be formed elsewhere, and then transported to

Fig. 2.4   63 Lincoln's Inn Fields, London, with Hyatt reinforced concrete floors (see Fig. 2.3) and walls of mass concrete (photo/copyright Lawrence Hurst).

its required location in the structure. Such work is known as 'precast' concrete. The traditional material for formwork is timber, which is cheap and easy to work, although it has limited scope for re-use as it is often damaged when the formwork is 'struck' or removed.

When reading the records of early concrete construction, it is remarkable just how early many of today's so-called 'modern' construction methods were already in use. For instance, metal formwork, which could be used many times and therefore reduced the cost and increased the speed of construction. Various inventors devised and patented 'concrete-building apparatuses' for use by house-builders and these were well reported in the press and, apparently, quite widely used. Other early examples include:

- 1830s precast concrete blocks;
- 1840s re-usable timber formwork for walls;
- 1840s continuous concrete manufacture and placement;
- 1870s larger precast concrete elements;
- 1870s climbing formwork for walls that could be repositioned without dismantling and reassembly.

What is most remarkable, however, is that virtually all the concrete in the nineteenth century was made and placed by human power. Cranes were rare, and hoists were human-powered. This makes many of the massive Victorian construction projects even more astonishing.

Design

There are three main components to the design of concrete structures—the concrete mix design, the choice of external dimensions to the concrete, and the location, number and size of any steel reinforcement inside.

Until almost the end of the nineteenth century, the first two were done on the basis of proportions ('rules of thumb') developed by long-established practice and embracing adequate safety; the third was largely not relevant until the 1890s, and is discussed in the next section. Having said this, by far the largest use of concrete was in compression, and typical crushing strengths of concrete were well known and measured when necessary. This figure alone would help ensure suitable dimensions for mass-concrete foundations.

The early days of reinforced concrete: from the 1890s to the First World War

The use of concrete

Context

Britain at the end of the nineteenth century was a major industrial and imperial nation, and its expansion continued during and after the Edwardian period until abruptly checked by the First World War. This expansion created considerable demand for more building in many areas of activity. Established and new industries were growing and demanding larger factories, warehousing for their raw materials and products, and offices for their administrative and clerical staff. The role of local authorities was expanding alongside that of private enterprise, operating public utilities such as electricity generation, telephone systems and tramways that needed power stations, exchanges and depots. An element of civic pride, coupled with civic prosperity, encouraged the commissioning of many imposing public buildings. The railways were still the dominant means of transport for goods and passengers and, like the country's seaports, needed more and more buildings to accommodate the consequences of their own success. In addition, motor vehicles, although in their infancy, were already creating a demand for new and wider roads and bridges. In order to respond to these demands for new construction, the building industry was open to exploiting new materials and techniques.

Experience of the loadbearing and spanning capabilities of cast and wrought iron had shown architects and their clients that buildings could be built higher with iron (more economically than with brick or stone), and with longer floor spans beyond the capabilities of timber. Two more new materials had been, in principle, available since the 1850s: steel and reinforced concrete. Neither, however, was widely used for construction until near the end of the century. In the case of steel, the reasons for this delay are reasonably clear with hindsight. First, there were technical difficulties in making steel successfully using the phosphatic ores most widely and cheaply available, and these were not overcome until the late 1870s. Secondly, steel was in greatest demand from other quarters, particularly to replace wrought iron in railway tracks, where its greater strength and toughness was quickly appreciated. There was also the natural (and not uncommon) conservatism of the British building industry. However, it is harder to understand the slowness in adopting reinforced concrete, not least because one of the pioneering patents for the concept was taken out in 1854 by W.B. Wilkinson of Newcastle upon Tyne. Britain was not alone in its dilatory approach to this new material but other nations, notably France, Germany and the United States, overcame their hesitancy and soon took the lead.

The first reinforced concrete structure in Britain was Weaver's Mill, a grain mill erected in Swansea in 1897 and regrettably, but perhaps unsurprisingly given its appearance, demolished in 1984 (Fig. 2.5) (Twelvetrees, 1907). The design was attributed to H.C. Portsmouth, although the structural design and construction were carried out by the French concrete specialist François Hennebique and his South Wales agent L.G. Mouchel.

Use of the material matured rapidly, so that by 1909 the Royal Liver Building at Liverpool's Pierhead, with its Liver Bird 94 m above street level, displaced the Ingalls Building of 1902 in Cincinnati, Ohio (64 m), as the world's tallest concrete building. In 1914, however, the outbreak of war put aside the civil use of concrete for 4 years. Thereafter, efforts were centred on military works, including Western Front defensive installations such as pillboxes, for which and from which much was learnt about

Fig. 2.5   Weaver's Mill, Swansea, near the end of its life (courtesy J.W. Figg).

the blast-resistance of concrete and the beneficial effects of reinforcement. Experimental concrete ships were also built at this time.

Building control regulations were slow to acknowledge the new material. No guidance on its use was given in the 1909 London Building Act (HMSO, 1909), the so-called 'steel frame act', that provided codified guidance on the design of steel, cast iron and wrought iron structures for the first time in Britain. However the Act made provision for the introduction of subsequent regulations, which eventually arrived in 1915 (London County Council, 1915). Before this, design was largely carried out by engineers working for the proprietary concrete firms, using texts written by the pioneers, and aided by the recommendations of committees set up by the Royal Institute of British Architects in 1907 and 1911 and the Institution of Civil Engineers, also in 1911 (Bussell, 1996a, b; Witten, 1996).

Architecture

The Victorian period had seen lively and often intense argument over architectural styles. Neo-Classical had its champions, as too had Gothic Revival and the Arts and Crafts Movement. Each style used traditional building materials (stone, brick and timber), although architects had become familiar with the use of cast and wrought iron for structures where spans were large or loads were heavy. Sometimes this contrasted sharply with the 'architectural' element of the building. Perhaps the best-known example of such a clash is at St Pancras Station in London, where the great vaulting single span of Barlow's train shed roof, with its wrought iron ribs, nudges against the back of the Midland Grand Hotel, Gilbert Scott's brick Gothic fantasia.

Nevertheless, some architects both welcomed and were intrigued by the new structural materials. Thomas Hardy the novelist, an architect in his younger days, would sketch the cast iron column capitals and wrought iron brackets of a railway station canopy while waiting for a train, and annotate his drawings with practical notes for future inspiration (Beatty, 1966). Richard Norman Shaw was interested in concrete and employed it in a number of his buildings (Saint, 1976).

In the early years of the twentieth century, a number of influential architects took an active interest in reinforced concrete. Not only did they use it in their own projects but, through their informed and enthusiastic accounts of such work coupled with membership of influential committees, they helped to establish it as a 'respectable' structural material that could be used with confidence. Two notable figures serve to illustrate this: Edwin O. Sachs and Sir Henry Tanner.

Sachs was a remarkable polymath, an 'architect, stagehand, engineer and fireman' (Hurst, 1998). His interest in theatres and his concern about the devastating effect of fires in buildings led him to found the British Fire Prevention Committee in 1897. This body undertook fire tests on construction elements and materials, and established the first European fire-testing laboratory in the garden of a house in London (Hurst, 1998). Its aim was to encourage the prevention of fire by all means possible, and so it is not surprising that Sachs should have been an early advocate of concrete, which (unlike exposed structural iron and steel) is inherently fire-resistant. In 1906 he established, published, edited and even, for a while, funded the journal *Concrete and Constructional Engineering*, which for 60 years was to cover developments and new construction in concrete. Sachs was a major force behind the establishment of the Concrete Institute in 1908 (Witten, 1996), becoming its first Council chairman. The Institute was founded to encourage the more informed use of concrete and reinforced concrete; in 1923 it became the Institution of Structural Engineers.

Sir Henry Tanner was appointed Chief Architect to Her Majesty's Office of Works in 1898 (Gray, 1988). He became aware of the economic potential of reinforced concrete and encouraged its use in Government projects. Tanner served on the 1907 and 1911 RIBA Joint Committees on Reinforced Concrete, and also on the 1911 Institution of Civil Engineers Reinforced Concrete Committee. In 1910 he was President of the Concrete Institute.

His patronage of the new material did much to encourage other architects to use reinforced concrete (Port, 1995). As there was no codified view on the design and use of reinforced concrete at the time, local authorities (and particularly district surveyors in central London) were chary about accepting proposals for its use. However, since Government buildings were Crown property they were exempt from local authority building control, and so Tanner was free to use the new material. His sorting office at the King Edward Street building of the General Post Office in St Martin's le Grand (1907–1910) in the City of London was among the largest reinforced concrete buildings yet built in Britain, and the first in central London (Fig. 2.6). It employed long-span haunched beams with

Fig. 2.6 King Edward Street Sorting Office in the City of London, during demolition (photo Michael Bussell).

generous storey-heights and floor-loading capacity, with a deep floor plate fed with daylight by lightwells—an early precursor of today's office block with open-plan layout and atrium (Twelvetrees, 1907). Like Weaver's Mill, this important early concrete building survived for nearly a century. Regrettably, it was demolished in 1998.

It has to be said that in the early years of this century, unlike in the 1920s and 1930s described below, architects were not yet seeing concrete as a material to be expressed or externally exposed in their buildings. The concrete would be exposed internally within functional buildings, or perhaps painted, but externally it was usually concealed behind a decorative facing of ashlar stone, brick, tiles, mosaic or render. In contrast, purely functional or engineered constructions such as bridges, water towers and jetties were usually left unclad.

The majority of the reinforced concrete buildings of the 1900s were indeed functional. A list of projects included in an early book reviewing achievements in concrete structures (Twelvetrees, 1907) runs to 15 pages, of which buildings and roofs occupy only a third. Most of these are warehouses, works and other utilitarian structures commissioned by local authorities, railway and dock companies, and the like. Architectural input to these would have been very limited. However, some notable buildings were designed by architects exploiting the potential of concrete. One such is the office block at Lion Chambers in Hope Street, Glasgow, by Salmon, Son and Gillespie, which is nine storeys high and was built in 1906 using the Hennebique system (Twelvetrees, 1907). The ability of reinforced concrete to cantilever was cleverly exploited to reduce the building's footprint at street level, and again by corbelling at fourth-floor level. The entire structure, including external walls, was formed in reinforced concrete. It was praised by Twelvetrees as 'the first example in this country of good architectural design realised in a building constructed entirely in concrete-steel' (a contemporary term for reinforced concrete; ferroconcrete was another). Nevertheless, Twelvetrees noted that 'all exterior surfaces were rendered with Portland cement mortar after completion, and great care was taken to secure an even surface'.

Structure

Unreinforced or mass concrete continued in use, particularly for railway works where thick, gravity-retaining walls were a simple and cheap way of upholding the soil at either side of cuttings. Unreinforced arch bridges made effective use of the compressive strength of plain concrete, echoing the traditional masonry arch. Such construction was applied in a spectacular fashion to viaducts on the West Highland Railway in Scotland, including the 1897 Glenfinnan viaduct with twenty-one 15-m spans constructed by 'Concrete Bob' (later to be Sir Robert) McAlpine. It is perhaps fortunate that reinforcement was not used in such an exposed location, so that the structure has not suffered corrosion damage, unlike many of its reinforced contemporaries.

In buildings, a notable structural use of unreinforced concrete is the nave roof of Westminster Cathedral, on which work started in 1895. The architect, J.F. Bentley, apparently without professional structural engineering advice, proportioned the four shallow square unreinforced concrete domes. The vaulted concrete is relatively thick, and lateral thrusts are resisted by the massive brick nave structure. Bentley claimed that he had shown that iron was not necessary for large spans (Gray, 1988).

Early use of reinforced concrete tended to follow the forms established by its predecessor, as is common when a new structural material becomes available. In this case, iron and steel framing provided the precedent. Thus, floors and flat roofs were made up of bays supported by an orthogonal grid of downstanding primary and secondary beams on columns. Taken to the extreme, the beam ends were haunched locally, mimicking the seating cleats used to connect an iron or steel beam to its stanchion (Fig. 2.7).

Fig. 2.7 Co-operative Wholesale Society warehouse, Quayside, Newcastle upon Tyne, 1897: detail of reinforced concrete beam–column joint, more akin to that found in an iron or steel frame (photo Michael Bussell).

Although initially slow to be adopted, reinforced concrete quickly became popular from the turn of the century. Many proprietary systems were employed, the commonest of which are described below. Here it may be sufficient to note that the French Hennebique system, one of the most successful, had first been used for buildings in Brussels and Tourcoing, France, in 1894–1895 (Delhumeau, 1992; Newby, 1996). By 1899, 3061 projects using the Hennebique system had been built worldwide, a figure that had grown to nearly 20,000 in 1909, by which time the company had 62 offices around the globe. By 1911, 1073 Hennebique projects had been constructed in Britain, including virtually all the earliest reinforced concrete bridges (Cusack, 1984; Chrimes, 1996a).

An alternative to the reinforced concrete slab was 'joist-concrete' or, as it was later known, filler-joist construction. This was a development from the fireproof floors of the nineteenth century described above, in which iron, and later steel, beams were placed at intervals and the gaps between were infilled with concrete to complete the floor. This technique also provided a degree of fire-protection to the beams themselves. Tables published by structural steel firms allowed architects and engineers to select suitable sizes of beams and slabs to suit spans and loadings.

Developments in structural concrete abroad attracted little attention in Britain until the 1920s. In the USA, Ernest Ransome had developed a ribbed concrete floor slab as early as 1880, and O.W. Norcross (1902) and C.A.P. Turner (1908) demonstrated flat slab construction that dispensed with downstanding beams. The concentrated shearing forces were transferred from the slab into the columns through locally thickened drop panels and enlarged column heads. Apart from the visual and practical benefits of an uninterrupted, flush soffit, the flat slab allowed freedom in column positioning, as these were no longer tied to a skeletal beam grid. Robert Maillart, in Switzerland, developed a more efficient form of flat slab in the same period. He demonstrated that only two orthogonal layers of reinforcement were needed to span the slab between the columns, whereas the American approach used a total of four layers, with the third and fourth layers orientated at 45° to the first two. Maillart also pioneered the 'flat arch' in 1901, which was effectively a beam with a concave soffit, producing an elegance that is widely acknowledged, and which anticipated his later and more daring bridges. The earliest flat slab construction in Britain is probably the Bryant & May factory in Liverpool, designed in 1919 by the Swedish engineer Sven Bylander (Addis in Stratton, 1997).

Novel reinforcing systems for both concrete and masonry were applied by the Frenchman Paul Cottancin. For concrete he developed the use of 'woven mesh'—an interlocking grid of wires or small rods that provided continuous reinforcement to walls and slabs, and indeed columns and beams. This is the precursor of today's welded sheets of steel fabric. His patent for this system was taken out in 1889, 3 years earlier than Hennebique's patent. For masonry, he placed vertical bars in the voids of perforated bricks and horizontal wires in the bed-joints, again anticipating more recent efforts to enhance the strength of masonry, especially when subject to bending. The Methodist church in Sidwell Street, Exeter, survives as the most notable example of his work in Britain (Edgell, 1985).

Concrete construction and design

Materials

Developments in concrete focused on the basic ingredients—a binder (typically Portland cement) that would harden when mixed with water, a cheaper filler or aggregate that would bulk out the mix, and the reinforcement.

Cement

Cement manufacture in the nineteenth century has been reviewed above. Towards the end of the century it was being produced by a diversity of companies, but in 1900 a major unification took place with the formation of The Associated Portland Cement Manufacturers (1900) Ltd, or APCM. This brought together no less than 24 companies with 35 cement works, mostly sited near the River Thames and the Kentish River Medway. Here, the raw materials for Portland cement (clay and chalk or limestone) were found in abundance, and the water and rail links provided cheap transport for the incoming coal to fuel the kilns and for the outgoing cement. It is doubtless no coincidence that the first chairman of APCM, Frederick White, was a close friend of the architect Richard Norman Shaw, whose active interest in concrete as a building material has already been noted (Francis, 1977).

A further step towards standardising cement and ensuring its quality was the publication in 1904 of a British Standard for Portland cement, later to become BS 12 (Engineering Standards Committee, 1904). This included requirements for specific gravity, setting time, fineness of grinding, soundness and tensile strength. At this time, cement was all of the 'ordinary' variety; rapid-hardening and sulphate-resisting cements were to come later.

Aggregates

In the earliest days of concrete it had been realised that varied sizes of aggregate were needed to ensure that the resulting concrete would be dense and strong. As nowadays, aggregate was usually categorised as coarse or fine, with size limits typically of $\frac{3}{4}$–$\frac{3}{16}$ inches (20–5 mm) for the coarse element, and below $\frac{3}{16}$ inches (5 mm) for the fine. There was also recognition that the aggregate should be clean and sound to avoid contamination or degradation of the concrete.

An early text offers a typical good practice specification of the time (Marsh and Dunn, 1908). Coarse aggregate could be 'shingle or broken stone', and washed as necessary to achieve cleanness and freedom from acid or alkaline pollution, as also was sand for use as fine aggregate. The damaging effect of fire on combustible or weak aggregates also had to be taken into account in choosing materials.

Some industrial waste products were countenanced for use as aggregate: 'furnace ashes or coke breeze', 'free from unburnt coal and thoroughly burnt', and 'as free as possible from sulphur and other impurities'. Steel reinforcement used in concrete made with these materials was to be protected by a coating of cement grout before the concrete was placed in order to reduce the potential for corrosion or chemical attack of the metal. It was tempting to use such waste material, which was obviously cheaper than natural materials and abundantly available. Indeed, broken brick and other less-wholesome waste materials are often found when examining early concrete structures. Particular concerns arise with breeze (a term derived from 'live coals'), clinker and other carbonaceous material. Firstly, these often retain compounds of sulphur and other aggressive chemicals that can accelerate corrosion of the reinforcement. This is a particular hazard when the concrete becomes wet, as can happen on poorly maintained flat roofs, in kitchens and bathrooms, and so on. Secondly, unburnt coal in the aggregate can support combustion in a fire, so that nominally 'fireproof' floors can be anything but fireproof.

The aggravation of fire hazard presented by unsuitable aggregates was recognised early on, and accordingly a Special Commission on Concrete Aggregates was formed in 1906 by the British Fire Prevention Committee. It set out to report on 'the aggregates suitable for concrete floors intended to be fire resisting, having due regard to the question of strength, expansion and the chemical constituents and changes of the aggregates' (Witten, 1996).

Reinforcement

Early reinforced concrete construction was dominated by specialist companies, usually with the benefit of patent protection of their system and, especially, of their reinforcement (Fig. 2.8). These companies either designed and built the work themselves, or licensed others who would be carefully selected and trained in order that standards were maintained.

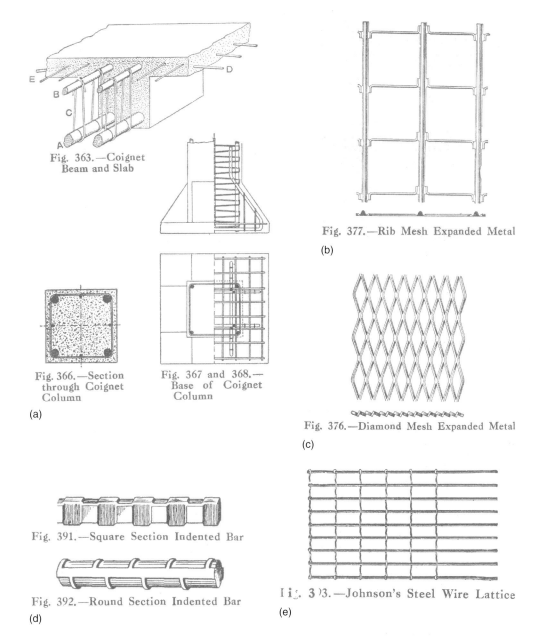

Fig. 363.—Coignet Beam and Slab

Fig. 366.—Section through Coignet Column

Fig. 367 and 368.— Base of Coignet Column

(a)

Fig. 377.—Rib Mesh Expanded Metal

(b)

Fig. 376.—Diamond Mesh Expanded Metal

(c)

Fig. 391.—Square Section Indented Bar

Fig. 392.—Round Section Indented Bar

(d)

Fig. 3?3.—Johnson's Steel Wire Lattice

(e)

Fig. 2.8 Some of the more common systems of reinforcement that may be encountered in buildings of this period (from Jones, 1920).

The earliest beginnings of reinforced concrete by W.B. Wilkinson and the American Thaddeus Hyatt were mentioned above. Others involved in studying, developing and testing this new material in the second half of the nineteenth century included the Frenchmen François Coignet, his son Edmond and Joseph Monier, and E.L. Ransome, an émigré Briton who achieved fame in the USA (Hamilton, 1956; Guillerme, 1986; Simmonet, 1992a). This work was largely ignored in Britain, where the practical use of reinforced concrete did not begin until the 1890s.

In 1892, François Hennebique, a French contractor, obtained a British patent for his system. This was the first system to be used in Britain, and was soon followed by others. All originated abroad, so that Britain was for once on the receiving end of a 'technology transfer'. Many systems were developed and documented, although in practice a few soon took the lion's share of the available work. The two leading systems were those of Hennebique (through his British agent L.G. Mouchel) and the American Kahn system.

In order to secure a patent, a company's reinforcement had to be distinctively different from existing profiles and to incorporate particular features. Hennebique was fortunate—or wise—enough to adopt the cheap, readily available round, plain, mild-steel bar. Its ends were fish-tailed to ensure anchorage of the bar within the concrete (essential if the steel and concrete were to work together) (Figs. 2.9 and 2.11). These bars provided the tensile resistance in beams and slabs, and augmented the compressive strength of the concrete in columns and walls. Shearing resistance in beams was provided by links of mild-steel flat strip, typically U-shaped with bobbed ends, wrapped around the tension bars and embedded in the compression zone of the beam. Column bars were restrained against buckling by wire strips. A later development was to bend up the main bars in beams to enhance resistance to shear.

This system of reinforcement has largely survived in common use to the present day. The only significant developments have been the supplanting of strips as beam links by bars, and the widespread adoption of ribbed or twisted bars whose deformed profile gives a better bond and anchorage within the concrete. These are either hot-rolled or cold-worked to give a higher yield strength than mild steel. Such bars were pioneered by Ransome in the USA in an 1884 patent.

The Kahn bar of 1902–1903 was different. It was based on a square section with 'wings' rolled onto two diagonally opposite corners (Fig. 2.10). These were slit so that they could be bent up diagonally, in short lengths, to be anchored in the compression zone of the concrete. The result was seen as a truss, with the square bar section forming the bottom boom and the compressed concrete the convex top boom, linked by the bent-up wings. Tests confirmed that these links were effective, and the concept

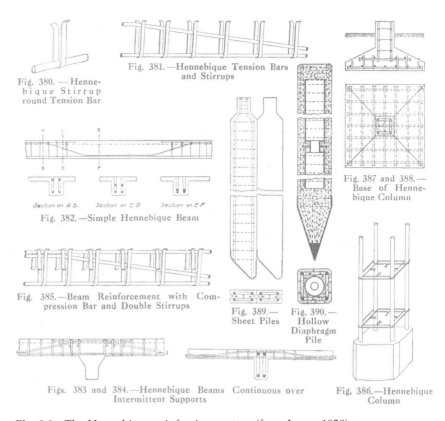

Fig. 380. — Henne-
bique Stirrup
round Tension Bar

Fig. 381.—Hennebique Tension Bars
and Stirrups

Section on A·B.   *Section on C·D*   *Section on E·F*

Fig. 382.—Simple Hennebique Beam

Fig. 385.—Beam Reinforcement with Compression Bar and Double Stirrups

Fig. 389.—
Sheet Piles

Fig. 390.—
Hollow
Diaphragm
Pile

Fig. 387 and 388.—
Base of Henne-
bique Column

Figs. 383 and 384.—Hennebique Beams Continuous over
Intermittent Supports

Fig. 386.—Hennebique
Column

Fig. 2.9   The Hennebique reinforcing system (from Jones, 1920).

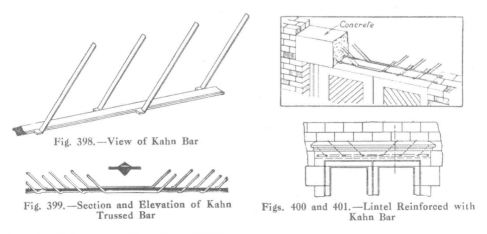

Fig. 398.—View of Kahn Bar

Fig. 399.—Section and Elevation of Kahn
Trussed Bar

Concrete

Figs. 400 and 401.—Lintel Reinforced with
Kahn Bar

Fig. 2.10   Kahn system (from Jones, 1920).

was embodied in the firm's name, the Trussed Concrete Steel Company, later known as Truscon.

Finally, it should be mentioned that other forms of steel were also used in reinforced concrete construction. Expanded metal (known in the USA as 'slashed metal') was introduced in the 1890s and found wide use. Structural steel sections such as I-beams were commonly embedded in concrete as foundation grillages for both steel- and concrete-framed buildings. For purely functional elements, particularly in wartime when steel was scarce, any steel to hand might be pressed into service, including conduit, second-hand cable, gas-pipes and fence posts.

Concrete mix design

Concrete was customarily batched by volume, or in the case of cement, by reference to weighed bags. A Mouchel specification of 1917 for Hennebique work prescribes 6 hundredweight of cement (305 kg) to 13.5 cubic feet (0.38 m³) of sand and 27 cubic feet (0.76 m³) of coarse aggregate, to give a mix with proportions of about 1 : 2 : 4 and a 28-day cube strength of about 15–20 N/mm² (Mouchel, 1917). The dry materials would be mixed manually on a clean surface using shovels or in a mechanical mixer.

Water was added to the dry mix. The amount of water should clearly be enough to react with all the cement to produce the chemical set that hardens it, and specifications at the time stressed the importance of not over-wetting the mix. This was mainly to avoid segregation of the mix when the heavier, coarse aggregate sinks to the bottom and the result is distinctly non-homogeneous concrete with variable and poor strength. It was not until 1918 that Duff Abrams, in the USA, published research showing that the strength of concrete was precisely dependent on the water–cement ratio: the stronger the concrete, the lower the water–cement ratio.

More recently, experience has shown that the durability of the concrete is related to the cement content. This is of particular importance in external and exposed locations, although in early concrete construction, strength was always the design criterion, not durability.

Construction methods

The essence of in situ concrete construction was already widely adopted: a wet mix of aggregate and cement or other binder was placed into previously prepared forms or, for groundworks, often poured straight into the excavation. The formwork ought to be clean, rigid and secure. Just as today, the formwork and its supporting falsework were often major works of timber engineering.

Reinforcement, if required, was to be clean, and cut and bent before being placed into the forms prior to concreting. As such, the process might appear to be similar to what is practised today, and in many respects that is true. However, there are significant differences, as will be shown.

Placing, compaction and treatment of concrete

A further factor in judging the amount of water was the need to ensure that the newly mixed concrete filled the forms as it was poured. Pragmatically, it was recognised that to produce a sound concrete, it should be as dense and as free of air as could be achieved. In the early years of the century, the methods of delivery available were a chute or wheelbarrow to transport the concrete; it would be lifted to its required level by a hoist and poured it into the forms. Here, it would be worked into position by shovel and spade, and then hand-compacted (maybe) using rammers or 'punning' tools. To assist compaction, especially with congested reinforcement, it was common to add water to the mix so that it flowed more easily and became, to some extent, self-compacting. Even so, concrete might not be able to penetrate crowded reinforcement, as shown in Fig. 2.11 in a photograph taken during demolition. In 1917, Eugène

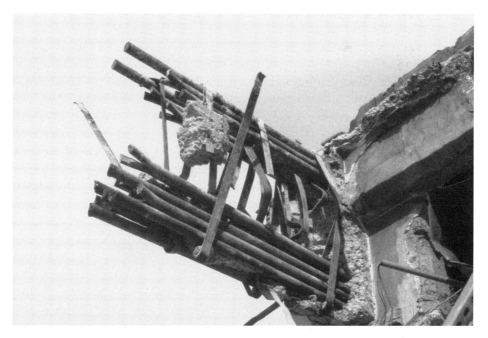

Fig. 2.11 King Edward Street Sorting Office. Column head with congested Hennebique main steel, exposed during demolition (photo Michael Bussell).

Freyssinet showed that mechanical compaction was superior, but its adoption was slow.

At this time, however, little thought was given to curing the concrete. Indeed one early specification recommended the early removal of formwork as 'the airing of concrete [sic] assists the time of setting to a remarkable degree and should always be encouraged' (Marsh and Dunn, 1908).

Load testing

Load testing of selected floor bays was frequently specified, and implemented as a form of quality assurance. Typically, the floor structure was to deflect no more than 1/600th of the span under 1.5 times the design-imposed load. This was not as onerous as it sounds, as design stresses were low and the typical floor slab was supported by beams on all four sides, so that it would behave more as a shallow vault in compression than as a slab subject to bending. Bridges were tested using railway locomotives coupled together, or a gathering of steam-rollers.

Precasting

Precasting was soon seen to be a useful technique that encouraged consistent standards, particularly for architectural concrete, and reduced the amount of site work (Morris, 1978). The architectural panels by Lascelles have already been cited as a late nineteenth century example. In the civil engineering field, early use was made of precast driven piles as a rapid and efficient foundation solution, particularly in poor ground and for marine works (Chrimes, 1996b). Hennebique obtained a patent for precast piles in 1897, and others were also quick to adopt this form.

Concrete blocks

Hand-operated concrete block-making machines were in use by the end of the nineteenth century and offered a cheap alternative to clay bricks, particularly where appearance was not a major issue. Comprehensive advice on their manufacture and use was readily available (e.g. Marsh and Dunn, 1908).

Procurement

Procurement of concrete construction in this period was often on the basis of what is now called 'design and build'. The specialist concrete firm would be responsible for the design and the construction of the concrete works. It would act either as the main contractor (commonly for engineering works such as jetties, bridges and water towers) or under appointment to the main contractor, which was more common in the case of architect-designed buildings. These specialist firms had a vested interest in maintaining the reputation of concrete in general, and of their own system in particular. Consequently, they would take great pains to ensure the quality of their work. In the case of Weaver's Mill, in Swansea, the first reinforced concrete structure in Britain, nothing was left to chance. The design was carried out by Hennebique's office in Nantes, in Western France, and also from here came not only the cement, aggregate and reinforcement, but also the labour and supervision. Only the water for the concrete mix originated in Swansea!

Design

The viability of reinforced concrete as a composite structural material is due to three factors:

- it is strong in tension and supplements the concrete, which is relatively strong when compressed but weak in tension and when sheared or twisted;
- the reinforcement bonds to the adjacent concrete, or it can be profiled so that it becomes mechanically anchored to the concrete, so that the two materials work together; without such a bond or anchorage, the loaded concrete beam would fracture as the reinforcement slides relative to the adjacent concrete;
- the reinforcement and most concretes have similar coefficients of thermal expansion (about $1 \times 10^{-5}/°C$) and so the bond and anchorage are not significantly affected by temperature changes in normal use.

Between 1877 and 1902, theoretical work and tests established the principles of reinforced concrete analysis (Hyatt, 1877; Wayss, 1887; Coignet and de Tédesco, 1894; Christophe, 1902). The reinforcement is designed to carry all tensile forces. Under a working load, at least, composite behaviour means that the strain at any point is the same for both materials, while the stresses in each are proportional to their relative elastic moduli.

From these assumptions, the stresses in the concrete and the reinforcement can be calculated for any given loading, and compared with limiting values of permissible stresses to produce a sound design. This so-called elastic modular ratio theory was in use for strength design until the 1960s, and is still used in today's limit-state codes to assess deflections and crack widths in serviceability calculations.

As noted above, there had been no statutory regulations for concrete design until the 1915 Reinforced Concrete Regulations were introduced in London (London County Council, 1915). Prior to this, design was largely undertaken by the specialist firms such as Hennebique, using methods derived from the works of Wayss, Coignet and de Tédesco, and Christophe, as noted above. These were published outside Britain, in French or German. The first British textbook for designers and constructors was published in 1904 (Marsh, 1904; Bussell, 1996b).

Technical information and exchange of experience started to become more widely available to designers and builders with the appearance of the publications already noted. Of particular benefit were the reports of various committees constituted by the Royal Institute of British Architects and subsequently the Institution of Civil Engineers (Joint Committee on Reinforced Concrete, 1907, 1911; Committee on Reinforced Concrete, 1910). The publication and discussion of examples of concrete construction were aided by the launch of *Concrete and Constructional Engineering* in 1906, and the foundation of the Concrete Institute in 1908 (Witten, 1996). In 1906, the weekly periodical *The Builders' Journal and Architectural Engineer* began publishing a monthly 'Concrete and Steel Supplement'. It offered 'articles by the foremost English, American and Continental specialists, theoretical and practical articles on the design and execution of Reinforced Concrete work, Special Systems of Reinforced Concrete Constructions, profusely illustrated with photographs and a number of Working drawings'. All this for a weekly outlay of two old pence (0.9p)!

A standard notation for reinforced concrete calculations was developed and eventually adopted from 1911. Design charts, graphs and formulae were used widely to save the time of busy designers working in a commercial environment and bearing in mind that the 'computer' of the day was a slide rule (Marsh and Dunn, 1908).

Between the wars: 1920–1945

The use of concrete

Context

By 1920, concrete was well established as a viable alternative to steel for most building types in Britain. However, the contracting industry was still rather biased towards steel since it was generally easier to organise on site, and because the industry tended to take its lead from the USA where steel was, by this time, dominant. The benefits of steel were well known:

- its high strength allowed wider spans between columns;
- simplicity of construction allowed speedy erection;
- it was easy to attach fixtures and machinery to steel frames, and to adapt the building as its use changed.

However, there were also three serious disadvantages:

- structural steel needed considerable fire-protection which, to begin with, was usually achieved by embedding the steel in concrete;
- an all-steel building would use more steel than one in reinforced concrete, a key issue in the post-war shortages of the early 1920s;
- there was nothing inherent in the use of steel that encouraged its expression as an architectural material in its own right, since steel-work was generally hidden beneath fire-protection, and building façades were usually of stone or brick.

Other factors also applied. Structural steel frames were essentially made up of 'stick' elements—beams and columns—which tended to encourage the use of orthogonal grids and planar elevations. More complex profiles increased the amount of steelwork needed and complicated member connections. In addition, steelwork usually required some protection against corrosion, especially if externally exposed or if built into the building envelope. Despite this there was a sanguine view that a modest thickness of masonry would suffice to weather-protect steel built into external walls. That this does not ensure effective protection against corrosion is a lesson now as familiar to architects and structural engineers as are the consequences of inadequate cover to reinforcement in concrete.

It was into this context that concrete began to emerge as a material with characteristics that would appeal to architects. Concrete could offer four utterly new opportunities:

- curved lines and surfaces;
- solid three-dimensional form;
- cantilevers;
- thin 'sheets' or structure (flat slabs, deep beams, shear walls, curved shells).

Individually, each these had first been used before 1920 in a wide variety of non-architectural applications such as factories, warehouses, maritime structures, airship hangars and even reinforced concrete boats. Sporadically they were discovered or reinvented by architects who gradually explored their potential during the 1920s (Collins, 1958). Finally, the several constituents finally coalesced, as it were, in the 1930s as the principal characteristics of what came to be known as the International Style.

In parallel with this movement towards the architectural expression of concrete, the inter-war period also saw the beginnings of industrialisation in construction. Systematic research was undertaken to establish how to improve concrete, i.e. how to control and achieve the desired structural properties of finished concrete, how to make and place it more rapidly, and how to achieve higher strength, quality, durability and reliability. This period also saw the early days of 'system building' that would flourish after the Second World War. In the 1920s, several systems were developed for making single houses of pre-cast concrete panels, but these required considerable investment and they declined as capital became increasingly scarce in the economic downturn at the end of the decade. In the late 1930s, the beginnings of a second phase of industrialisation was curtailed by the outbreak of war.

Architecture

At the risk of annoying some aficionados, it is not far from the truth to say that relatively little concrete architecture of note was executed in Britain during the 1920s. Some rare hints of what was to come in the next decade were the New Hall for the Royal Horticultural Society, London (1923–1926, architects J.M. Easton and H. Robertson, engineer Oscar Faber), the Palaces of Engineering and Industry for the British Empire Exhibition at Wembley (1924, architect Maxwell Ayrton, engineer Owen Williams) (Addis, 1997), and the wonderful but little-known grandstand at Northolt Park racetrack in west London, now lost (engineer Oscar Faber) (Figs. 2.12 and 2.13).

Among the first commercial buildings of the 1930s with an expressed concrete structure was the office of W.S. Crawford in High Holborn, London (1930), by Frederick Etchells (who had translated Le Corbusier's *Vers une architecture* (1923) into English in 1927). The new International Style inspired domestic architects, and many houses of reinforced concrete were built during the 1930s. Although concrete was still less common than steel as the choice for larger buildings, exceptions were made when the building was seen as needing to express a modern image, such as cinemas, rail and underground stations, bus stations and a new building type, the airport building. This is exemplified by the 1936 terminal for Gatwick Airport, south of London (architect Hoar, Marlow and Lovett) (Figs. 2.14 and 2.15).

During this era of architectural adventure, many engineers were also stimulated to innovate, not only in technical matters but also in changing their very role (Newby, 1996). A few important engineers realised that they had much to offer architects who generally were not well aware of the opportunities of the new building technologies. They found that they

Fig. 2.12   Grandstand at Northolt Park racetrack, front elevation (1929) (from *Concrete and Constructional Engineering* (1929) **24**: 742–744).

Fig. 2.13   Grandstand at Northolt Park racetrack, rear elevation (from *Concrete and Constructional Engineering* (1929) **24**: 742–744).

Fig. 2.14 Gatwick Airport terminal, exterior view (from *Concrete and Constructional Engineering* (1936) **31**).

Fig. 2.15 Gatwick Airport terminal, interior. The internal wall at first floor level, supporting the control tower and roof, consists of a circular Vierendeel girder supported at six points (from *Concrete and Constructional Engineering* (1936) **31**).

could make a direct contribution to the finished product, the architecture itself, by working in close collaboration with architects in the conception of buildings.

This new approach was best exemplified in the work of Owen Williams, without doubt the greatest British building engineer of the twentieth century. In fact, Williams went one step further and was appointed as architect to many of his projects, although he was not qualified as such. He worked predominantly in concrete, and made a number of striking contributions to the art of building (Cottam, 1986):

- the expression of the function of structural members by shaping them to reflect the stresses and bending they were carrying, for example, the columns and floors of the Daily Express building, London (1929–1931), Boots 'wets' and 'drys' buildings (1930–1932 and 1935–1956) (Figs. 2.16 and 2.17) (*Architects' Journal*, 1932, 1938, 1994, 1997), and the later roofs in the BOAC maintenance headquarters building, Heathrow Airport, London (1950–1955);
- the integration of services into his concrete structures, for example, the main columns of his Dorchester Hotel scheme, London (1929–

Fig. 2.16 Column detail in Boots packed 'wets' factory, Beeston, Nottingham, 1930–1932. Engineer and architect Owen Williams (photo Bill Addis).

Fig. 2.17   Column detail in Boots 'drys' factory, Beeston, Nottingham, 1935–1936. Engineer and architect Owen Williams (photo Bill Addis).

1930, completed by other architects) were a cluster of four columns with a central void to carry vertical service runs, and the Pioneer Health Centre, Peckham (1933–1935);

- the use of the curtain wall, for example the Daily Express building, London, and Boots 'wets' building, Nottingham (1930–1932) (Fig. 2.18);
- sheer structural boldness and ingenuity, for example one roof in the Boots 'drys' building (1935–1936) has 9-feet (2.75-m)-deep valley beams spanning 215 feet (65.6 m) on just two columns, suspended from an adjacent building at one end and with a 48-feet (14.6-m) cantilever at the other, far greater than ever before (and seldom exceeded since).

Ove Arup was another of the engineers who pioneered the collaboration with architects. His contribution has unjustifiably tended to over-

Fig. 2.18 Exterior view of Boots 'wets' factory, Beeston, Nottingham, 1939–1932. Engineer and architect Owen Williams (photo Cement and Concrete Association).

shadow that of Williams, due mainly to the subsequent activity and fame of the firm that carries his name. Arup worked for Christiani and Nielsen and for Kier, who were Danish contractors and very experienced in concrete, mainly on civil, industrial and maritime projects. Arup saw the benefits that his knowledge of construction could bring to architects, especially by adapting the rapid casting of concrete for use in buildings. He used highly simplified concrete profiles which were amenable to the use of climbing formwork that his firm had developed for grain silos. He first did this, working closely with Lubetkin and Tecton, on the gorilla house (1933) and the penguin pool (1934) at London Zoo, and then at the Highpoint I flats in Highgate, north London (1935) (Allan, 1992).

In this latter project Arup, like others before him, came up against the regulations. He proposed doing away with many of the columns that Lubetkin had originally included by thickening the walls sufficiently to carry all the vertical loads. Eventually, with some additional (unnecessary) reinforcement added as a compromise, his idea went through, but only because Highgate lay just outside the jurisdiction of the London County Council, for which District Surveyors were the respected, but not always flexible, arbiters of building regulations.

Structure

That the Highpoint I flats are still of interest for their structural novelty serves only to emphasise that the inter-war period was generally a time of consolidation rather than innovation in concrete construction (Newby, 1996; Bussell, 1996b). However, there are two notable exceptions: in structural form, the first examples of concrete shells, and in construction technology, the development of prestressing for concrete structures.

Shells

The first shells were 'bespoke' designs, tailored to the particular building and responding to its architectural concept. This compares with the post-war 'off-the-peg' standard shell forms, as described in the next section. A very early reinforced concrete shell was used for the dome of the Church of St John in Rochdale, Lancashire (architect Bower Norris, engineer Burnard Green) (Figs. 2.19 and 2.20). Reputedly the first real concrete shell was the roof of an aircraft hangar at Doncaster Aerodrome (Fig. 2.21).

Fig. 2.19 Exterior view of the Church of St John, Rochdale, 1925 (from *Concrete and Constructional Engineering* (1925) **20**: 540).

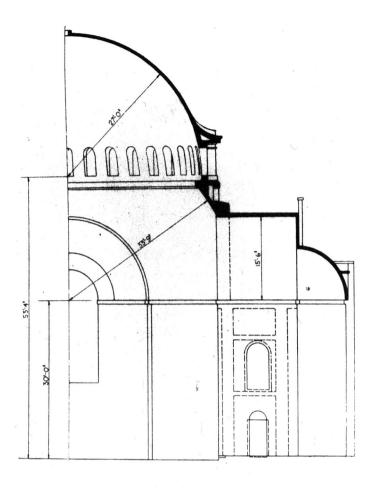

· SECTION B.B.

Fig. 2.20 Section through the Church of St John, Rochdale, 1925. At 54 ft. (16.5 m), this dome was, in fact, smaller than the largest of the domes at Westminster Cathedral (60 ft., 18.3 m), built by J. F. Bentley in 1906. That dome, however, had been of unreinforced concrete and was 13 in. (0.33 m) thick at the crown, increasing to 3 ft. (0.91 m) at the springing. The reinforced concrete-shell at St John had a constant thickness of about 8 in. (0.2 m) (from *Concrete and Constructional Engineering* (1925) **20**: 540).

Prestressed concrete

The first suggestion that prestressing might overcome the problem of concrete cracking under even modest tension loads was made in Germany before the First World War, but no practical means was devised for introducing the prestress, or for overcoming the inevitable movement and loss of prestress caused by creep and shrinkage in the concrete. Eugène

Fig. 2.21 Doncaster Aerodrome hangar, 1936. The concrete shell is approximately 4 in. (100 mm) thick. The building consisted of two bays, each 90 ft. × 30 ft. (27.5 m × 9.1 m). It is reputed to be the earliest genuine concrete shell in Britain (from *Concrete and Constructional Engineering* (1936) **31**).

Freyssinet solved these problems and patented his system in 1928. The early use of prestressing was in bridges and structures in contact with water, since the prestressing would minimise or eliminate cracking and hence help to keep moisture away from the steel reinforcement. A further essential ingredient that made prestressing a viable option was the development of high-yield steel wire in the late 1930s. High-tensile steel had been available for some time, but it had not been possible to control accurately the stress at which it began to deform permanently (the elastic limit or yield point). This was essential to ensure that the level of prestress could be guaranteed. Little use was made of prestressing in Britain until the Second World War, but a few examples have been documented (Andrew and Turner, 1985).

Concrete construction and design

Materials

After the first two decades of experimenting with concrete, the inter-war period was when firms really got down to developing a thorough understanding of the material and how it could be used most effectively and reliably. Until this time, mix design had been a relatively crude process using simple proportions by volume, and the resulting strength of the concrete could not be predicted with any great accuracy.

Material standards for concrete

The final years of this period saw the burgeoning of standards intended to ensure that materials, for both concrete and reinforcement, could be specified and supplied against agreed criteria. This was not an entirely new development; as noted earlier, the precursor of BS 12 for Portland cement appeared as early as 1904, and had been regularly revised as knowledge and experience grew. Reinforcement standards had also begun early with the issue of BS 15 (1906) initially covering mild steel which had largely displaced the early proprietary reinforcing systems, and later extended to cover hot-rolled steel (see below). Coarse and fine aggregates were dealt with in BS 882 in 1940.

Research into the characteristics and properties of concrete

During this inter-war period, a great deal of research—much of it in Britain, but more in the USA—was undertaken to improve concrete mix designs in order to enable the specifier to achieve with certainty the quality and performance required of a concrete mix for a particular use and environment. This was of great importance to both concrete contractors and the designers of concrete structures, although at this time there were very few independent consultants designing concrete structures. It should be added that most of these developments were of little direct interest to architects. This was the period when our modern understanding of the material was given its first solid foundations with a wealth of reliable information about:

- the effects of mix design;
- the long-term strength of concrete;

- the rate at which its strength develops over months and years;
- the influence of temperature on concrete making;
- its workability when freshly made, and as it cures and sets;
- the speed at which it sets, and the factors affecting this;
- the effects of contamination in the water used to make the concrete;
- the effects of air entrapment, and how to reduce this using vibrators;
- the effects of non-uniform mixing and separation of the mix into layers;
- the most effective ways of placing reinforcement to ensure that concrete fills the formwork without voids;
- the best means for easing the release of shuttering to prevent damage to the concrete;
- the relative benefits of different types of reinforcement bar, especially in order to prevent it debonding and pulling out of the concrete when stressed;
- the penetration of water, as liquid and vapour, into concrete, and its role in the corrosion of reinforcement.

Shrinkage and creep of concrete

One major contribution to understanding concrete was made during this period in Britain. This was the work by Oscar Faber and the Building Research Station in the late 1920s. They investigated the shrinkage that concrete undergoes when curing and subsequently, and also the phenomenon known as creep. Creep is the process by which concrete continues to deform to a limiting extent, albeit slowly, when loaded over time. It was the discovery of this process and its understanding that was the necessary precursor to developing the successful application of prestressing of concrete, which Freyssinet finally achieved in the late 1920s.

Lightweight concrete

Another development in the inter-war period was lightweight concrete, with a density of between 60% and 80% of ordinary concrete. This had been used since Roman times, when natural pumice stone or fired tiles were used as the aggregate instead of stone or gravel, but it had not yet been industrialised. A more scientific approach to creating lightweight concretes began in the 1930s in the USA using foamed slag, a waste material from the steel-making process, or expanded clay. Since the lower density results mainly from entrapping air in the concrete, lightweight concretes can only be used where a relatively low strength is needed. This is ideal for floors and the walls of low-rise buildings, including housing.

It also offers the benefit of improved thermal insulation properties, although this was not to become significant for several decades.

Material standards for reinforcement

Standards for reinforcement tended to lag behind the introduction and establishment of new grades and types of steel, as is common with all materials. Originally issued as a standard for structural steel, BS 15 had then specified ultimate tensile strength (UTS) but did not specify yield point. During the 1930s it came to be realised that the yield point was at least as useful and important as the UTS, since it provided a measure of the real 'factor of safety' should overstress occur. Accordingly, BS 15 was amended so that steel could be specified with or without a defined yield point, although it was later withdrawn following subsequent developments. These were reflected in the issue of BS 785 (1938) for hot-rolled bars and hard drawn-wire with three grades of strength, followed during the war by BS 1144 (1943) for cold-worked bars and BS 1221 (1945) for steel fabric, that is to say welded mesh. BS 1144 covered single twisted bars, in which the steel is twisted cold to form a barley-sugar profile, and also 'twin twisted' bars which comprise two such bars twisted together. The latter provided a heavier but more compact reinforcement section. BS 1221 also embraced expanded metal, which had been used since the late nineteenth century as reinforcement in concrete.

Construction methods

The thorough understanding of the properties of concrete not only led to its more reliable use, but also to the much more efficient making and placing of concrete. The first large batching and mixing plants were used in 1925, concrete pumping was first used successfully in 1927, and 1930 saw the first deliveries of ready-mixed concrete, albeit in small loads and in curious vehicles. The gunite process, by which concrete can be sprayed onto reinforcement or to effect repairs, was developed and brought into wide use in the mid-1920s.

Compaction methods, and their consequences

Freyssinet's study of the benefits of mechanical compaction, made in 1917, did not find rapid application, even by Freyssinet himself, who did not introduce mechanical vibrators until 1924. Before this, and indeed afterwards by less enlightened contractors, concrete mixes were made 'self-compacting' by the liberal use of water so that they would flow easily in the formwork. A limited compaction was sometimes achieved using hand tools to 'pun' the concrete in an effort to drive out entrapped air. This was almost literally a hit-or-miss practice, while the wet concrete mixes led to weaker concrete and a reduced durability.

Precasting

Research on concrete, as outlined above, had demonstrated the enormous benefits of being able to control the production of concrete precisely. One way to achieve this was to adopt the factory system, with pre-cast concrete panels and other components being made on a production-line basis. This could offer covered working conditions, training to carry out repetitive tasks properly, and close supervision and checking of the products. These could then be transported by lorry for assembly in situ. In the early 1920s there was a flurry of activity making housing in this way, but in Britain it died out largely through lack of investment.

On the other hand, precast floor components found a wider market amongst contractors who recognised their benefits. Typical of these components was a thin precast concrete plank, encasing reinforcement, which was itself part of a lattice that projected above the plank. The lattice acted as stiffening to a component that was light enough to be carried by two or three labourers, and placed at defined spacings onto walls or beams. Hollow fired-clay or precast concrete blocks would then be laid in place bearing onto adjacent planks, and sometimes with a mesh placed on top of the blocks. Concrete would be poured in situ, and bond to the lattices and to the planks to form a deeper composite concrete floor. Suitably designed and reinforced, such floors needed no propping and were quick to build, while the hollow blocks reduced the dead-weight of concrete in the tension zone of the floor where it was of little use. Such floors also found wide use in steel-framed construction.

Air-raid shelters

Arup and the architectural practice of Tecton, including Lubetkin, developed a construction technique for deep shelters that was not adopted at the time, but led towards today's 'top-down' construction techniques for forming deep basements economically and without using major temporary works to protect adjacent structures from damaging ground movements. They prepared a scheme for Finsbury Borough Council in the late 1930s, when war with Germany seemed unavoidable, based on deep concrete cylindrical shafts with spiral ramped floors. These were to be excavated from ground level, with walls and floors constructed in sections downwards as excavation proceeded, and with the floors bracing the walls to resist inward earth pressures. These were identified for post-war use as underground car parks—a valuable asset were they to exist today. However, the government of the time favoured other means of mass air-raid protection (Mallory and Ottar, 1973).

Design

Along with the growth of materials standards, this period saw the beginnings of serious 'codification' of design guidance and procedures for what might be termed 'routine' structures (Bussell, 1996b). The design of more challenging structural forms such as shells, folded plates or the spiral ramps at the London Zoo penguin pool was, and remains, less amenable to codification. Before computer programs made the analysis of such structures a relatively simple task, they were the province of the mathematically talented specialist designer.

The 1915 London County Council Reinforced Concrete Regulations were followed in 1933 by a 'code of practice for the use of reinforced concrete in buildings' drafted by a committee of the Building Research Board of the Department of Scientific and Industrial Research (Reinforced Concrete Structures Committee, 1933). Its coverage of structural elements went as far as to include flat slabs. The code also took what at first glance appears to be a welcome and supportive approach to what it called 'special forms of construction not otherwise provided for', which would be acceptable if their strength and durability were at least equivalent to what the code called for. How this was to be demonstrated imparts a Delphic ambiguity to this otherwise enlightened approach. A similar code was issued in 1939 (Building Industries National Council, 1939), but was little-used with the advent of war. Both of these codes were 'permissible stress' codes, following their predecessors back to the turn of the century.

Post-war reconstruction and expansion: 1945 to the mid-1960s

The use of concrete

Context

The construction environment in Britain in the immediate post-war period was dominated by a tremendous need for new construction. This arose from the urgency of replacing bomb-destroyed property, particularly housing, augmented by the new Labour government's commitment to building the Welfare State and other social reforms which required hospitals, new schools, similar public projects and the replacement of slum housing that had survived the bombs.

There was, however, a great shortage of steel, as it was rationed and expensive. Other building materials were also in short supply, in particular bricks for housing. A further difficulty was the militancy of the steel industry's trade unions. Strikes that interrupted the manufacture, delivery and erection of steel were common, and were catastrophic to many projects until well into the 1960s. These factors encouraged the use of concrete and the search for high-speed construction methods.

The drive towards industrialisation, that had had two false starts before the war, finally took place (Finnimore, 1989). The first area of development, perhaps less well-known and certainly less notorious than the later high-rise system-built blocks, was the production of houses, typically of two storeys. These often employed light 'frames' and roof trusses of precast concrete, steel or aluminium. One positive aspect of this programme was the endeavour to exploit both the capacity and the attitudes of the industrial expansion that had been applied to the war effort, now itself seeking new markets. Thus, for example, aircraft factories were diverted towards producing aluminium house frames, assisted by the Aircraft Industries Research Organisation for Housing (AIROH). Onto these AIROH frames were secured precast concrete floor panels and wall cladding panels of various sizes and materials. The pitched roofs were commonly clad with concrete tiles. A large variety of such low-rise housing systems were employed in the years after the war and, despite problems of spalling and corrosion in some, many of these houses have now acquired, after half a century, a quasi-traditional character.

During the 1950s and 1960s many contractors adopted, adapted, developed and patented their own systems for building 'homes for heroes' as the pre-war slums were replaced by high-rise blocks in nearly every city in Britain. However, of about 170 systems that were current in the mid-1960s, most were imported from abroad, especially from France, Scandinavia, Germany and the USA. Just as foreign ideas and firms dominated concrete construction at the turn of the century, so the pattern was repeated in the post-war era.

The fifties and sixties saw the growth of another development that had begun before the war—the creative collaboration of engineers and architects. Ove Arup is the consulting engineering firm perhaps most associated with this in Britain, but there were others: Owen Williams, Felix Samuely and Partners, Oscar Faber, Harris and Sutherland, Anthony Hunt Associates and numerous smaller practices. Some were launched by experienced engineers from the contracting or manufacturing side of the concrete industry, who then concentrated on a particular market. An example of this is the practice of Jan Bobrowski and Partners, designers of many notable precast grandstand structures; Bobrowski had been the chief engineer for a major precast concrete manufacturer.

Many of the works resulting from these collaborations were remarkable, representing giant leaps in understand and capability in all types of structure. The main legacy in the use of concrete was mastering the art of thin shells: domes, barrel vaults, hyperbolic paraboloids, folded plates and many variations on these basic types. For the first time since the mid-1800s, Britain was among the world leaders in construction. These structures met the needs of the time by providing large-span enclosures which exploited the compressive strength of concrete with only a modest requirement for scarce reinforcement. The formwork and falsework needed for these structures was substantial, but timber and the associated labour were relatively cheap and to hand. Today the economics are entirely altered.

A more equivocal development from the late 1950s was the commercial property boom, which led to entire quarters of town and city centres, usually of very modest architectural scale, being demolished and replaced by large buildings. A relaxation of planning regulations and improvements in fire-fighting technology combined to permit the dominating office block, often rising as a tower from a podium block. Public unhappiness with such large-scale schemes has surely been a factor in the subsequent support for heritage and conservation, matched by hostility to new buildings and the poignant and disparaging phrase 'concrete jungle'.

Architecture

Ove Arup had realised the potential of the simplified concrete construction methods he had devised with Tecton before the war. In a report written in 1944, he outlined the options for a variety of two-storey buildings using what he called the box frame idea. In his covering note, Arup said: 'this memorandum is a further development of my proposal to provide "safe housing in war time" on the principle of the "box frame" construction, published two years ago. It is to be hoped that the high resistance of the proposed structure against high-explosive bombs has now ceased to be of importance. Further investigations have, however, strengthened a conviction, which I expressed at the time, that this method would prove to be of value also in peace time' (Arup, 1944).

This same idea was expanded many times by Ove Arup and Partners in a number of blocks of flats, for example in Hallfield Estate, Paddington, London (1946–1954), and Spa Green Estate in Rosebery Avenue, London (1949–1950), both with architect Tecton (Figs. 2.22–2.25). This system came to be known as egg-box or crosswall construction.

Le Corbusier was a dominant influence on the young architects of the late 1940s and 1950s, and to them an obvious solution to our housing needs was to build upwards. Unfortunately, their architectural ideals

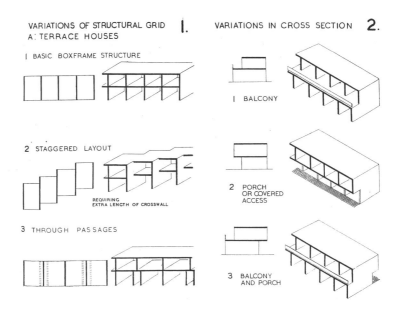

Fig. 2.22 Drawing from Ove Arup's report on box-frame construction (courtesy Ove Arup Partnership).

were rather lost in the rush to build as much and as quickly as possible. Systems proliferated as all the major contractors developed their own patent methods for high-rise flats and other building types (Fig. 2.26). One national survey reported no less than 170 proprietary building systems, mostly non-British, many of which were used in Britain under license (Diamant, 1964, 1965, 1968; Bussell, 1996b).

As is always the case with a new type of construction implemented in haste, too little time was spent on the design and planning of many of these systems in Britain. A host of problems arose. Some were a result of transplanting designs out of context. Technically and socially, the British climate was not the same as in Le Corbusier's Unité d'habitation in Marseilles. Poor design, construction and/or supervision were often followed by a lack of building management, inspection and maintenance. This led to many unpleasant consequences for the residents: water leaking from flat roofs, condensation on uninsulated external walls from gas cookers or heating systems, followed by mould growth, spalling concrete, detached claddings of tile or brick, faulty and vandalised lifts, and, not least, social problems associated with young children living high up.

The most notorious event associated with this era, which perhaps finally ensured an end to the high-rise system-building mania, was the

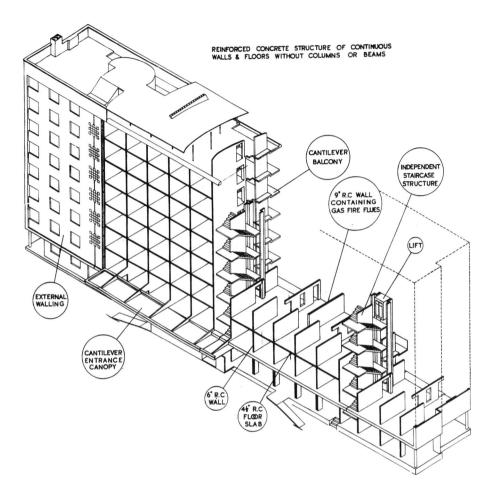

REINFORCED CONCRETE STRUCTURE OF CONTINUOUS
WALLS & FLOORS WITHOUT COLUMNS OR BEAMS

CANTILEVER BALCONY

INDEPENDENT STAIRCASE STRUCTURE

9" R.C WALL CONTAINING GAS FIRE FLUES

LIFT

EXTERNAL WALLING

CANTILEVER ENTRANCE CANOPY

6" R.C WALL

4½" R.C FLOOR SLAB

Fig. 2.23   Cutaway isometric of Spa Green Estate, Finsbury, London, 1949 (courtesy Ove Arup Partnership).

dramatic and fatal collapse in 1968 of one whole corner of Ronan Point, a 23-storey tower block in east London, following a relatively minor gas explosion. The investigation into this collapse (Griffiths *et al.*, 1968) led to new, explicitly stated requirements in building regulations and codes of practice for robustness. These aim to ensure that an 'event' such as a modest explosion will not lead to 'disproportionate' collapse, such as sadly occurred at Ronan Point. Recent experience of terrorist bombings in Britain and elsewhere, for example the Federal Government building in Oklahoma City, has highlighted the desirability of this approach.

Fig. 2.24 Spa Green Estate, Finsbury, London, 1949, seen under construction (photograph Sydney Newbery, courtesy Ove Arup Partnership).

Fig. 2.25 A completed building on the Spa Green Estate, Finsbury, London, 1949 (photograph Sydney Newbery, courtesy Ove Arup Partnership).

Fig. 2.26   The Taylor Woodrow–Anglian precast housing system, typical of many (from Deeson, 1964).

Structure

The concrete shell was undoubtedly the most dramatic type of new structure that this century has seen. As with many ideas, it began small and rather earlier than is usually realised. On the continent, shells appeared in the late 1920s, and Eduardo Torroja was an early exponent in Spain during the 1930s (Newby, 1996). As mentioned above, some thin shells had been built in Britain before the war, e.g. the Church of St. John in Rochdale (1925), at Doncaster aerodrome (1936) (see Figs. 2.19–2.21) and Chessington railway station, but these were on a scale altogether different from what came after the war.

The first major shell roof was the Brynmawr rubber factory in South Wales (1947–1950, architect Architects Co-Partnership, engineer Ove Arup and Partners). At the time of writing, this listed structure is being demolished, despite convincing arguments being put forward about its possible rehabilitation and re-use (Perry, 1994). It is noteworthy that from the point of view of the concrete structure, the building has been given a clean bill of health. Its fate is very much influenced by its location in a depressed post-industrial community that currently sees the derelict structure as a sign of economic decline and failure as tangible as rusting pithead gear at an abandoned coal mine. These issues are likely to apply to many other major concrete structures as their original purpose comes to an end, as evidenced by the fate of Weaver's Mill and the King Edward Street build-ing. Architectural or technologico-historical merit may not be enough to ensure survival.

Some other notable shells erected during this period were the bus garages at Bournemouth (1951) and Stockwell, London (1954, architect Adie Button and Partners, engineer A.E. Beer and Partners), the Bank of England printing works at Debden (1953, architects J.M. Easton and H. Robertson, engineer Ove Arup and Partners), and Smithfield Market (1962–1963, architect T.P. Bennett, engineer Ove Arup and Partners). Par-ticularly notable was the roof of the Commonwealth Institute in London (1962), one of the few large hyperbolic paraboloid roofs in Britain (archi-tect Robert Matthew and Johnson-Marshall and Partners, engineer A.J. and J.D. Harris) (Anchor, 1996; Morice and Tottenham, 1996). The enthu-siasm with which such applications of concrete were welcomed in the 1950s was remarkable, and can be captured only by contemporary accounts (de Maré, 1958).

On a smaller scale are the shell roofs that cover many hundreds of markets and factory buildings constructed up and down the land during the 1950s and early 1960s (Figs. 2.27–2.30). They were usually the work of specialist concrete firms, acting as design-and-build contractors to realise an architect's basic layout and scheme. Many of these roofs have lasted well even in our moist climate. As with other reinforced concrete buildings, their life depends initially on the quality of the site personnel and the care with which they checked the placing of the reinforcement, and the quality and placing of the concrete, as well as on the subsequent maintenance regime for the building. Maintenance remains the Cinderella of the building industry; it has yet to go to the ball.

Like many of the small, curved shell roofs, folded-plate structures in concrete were used in many small local authority, commercial and factory buildings and, in general, have gone unremarked. One charming example is a garage in Plymouth (1960, architect Roseveare, engineer Felix Samuely and Partners) (Fig. 2.31).

The use of prestressed concrete made little direct impact on mainstream architecture; it tended rather to be an engineer's elegant way of max-imising the efficiency with which the materials were used. As with other technical developments, a few individuals were prominent. Samuely, a particularly ingenious engineer, used prestressing in both steel (such as in the Skylon at the Festival of Britain, London 1951) and concrete (such

Fig. 2.27 A north-light, concrete-shell factory roof at Cadbury's factory, Moreton, Cheshire, under construction (courtesy Institution of Civil Engineers).

Fig. 2.28 Kidbrooke Comprehensive School, Blackheath, south-east London, 1949–1954, with a shell roof span of 113 ft. × 75 ft. (34.4 m × 22.9 m). Architect Slater Wren and Pike, engineer Ove Arup & Partners (courtesy Ove Arup Partnership).

Fig. 2.29   Pannier Market, Plymouth, 1960, exterior view. The north-light shell roof has six bays and covers a total area of 192 ft. × 76 ft. (58.6 m × 23.2 m). The market was originally conceived with two trading floors. The upper one was abandoned, but the roof was kept at the original height. The result is a truly splendid interior. The building won a Civic Trust Design Award in 1960. Architect Herbert Walls and Paul Pearn, engineer British Reinforced Concrete Engineering Co. (photo Bill Addis).

as at the Malago factory, Bristol, 1948), and he spread his ideas widely (Higgs, 1960). The late Alan Harris (subsequently a partner in Harris and Sutherland) had worked in Freyssinet's firm in France immediately after the war. He brought prestressed concrete back with him to Britain, and used it, for instance, at the British European Airways hangar at Heathrow Airport, and the Spekelands Road Freight Depot, Liverpool (Newby, 1996; Addis, 1997) (Figs. 2.32–2.35). In the immediate post-war period, prestressed concrete was given particular (and uncharacteristic) encouragement by the government because it enabled steel to be used more efficiently. The steel used in prestressed concrete was not subject to the same rationing that restricted the use of steel in conventional reinforced concrete and steel structures (Walley, 1996).

By the 1960s there were a great many prestressing systems available in Britain for beams and trusses, which were widely used in floor and roof structures. The 'Milbank' floor, developed by the Ministry of Works, comprised inverted pre-cast concrete troughs spanning between beams of concrete or steel. 'X-joists' were shallow prestressed beams between which were placed hollow concrete blocks. The 'Laingspan' system,

Fig. 2.30 Pannier Market, Plymouth, interior view (photo Bill Addis).

comprising shallow prestressed concrete trusses, was developed specifically for the massive school rebuilding programme of the 1950s (Walley, 1996) (Fig. 2.36). More recently, prestressed, precast concrete planks by firms such as Bison have proved very popular, especially for the speed with which they can establish a load-bearing floor as the building is erected (Fig. 2.37).

It is always difficult to differentiate between prestressed and ordinary reinforced concrete in slab structures; the difference may be apparent

Fig. 2.31 Folded plate roof over car showroom, Turnbull's Garage, Plymouth, 1960. The folded plate roof is supported on ten columns beneath the valley points. The adjacent flat circular roof covered the filling station area of the garage, which was the first self-service filling station in Britain (photo Bill Addis).

only in roof trusses, long-span beams, and mullions with member sizes suggestively more slender than normal. The main drive for developing the use of prestressed concrete was largely economic rather than any architectural benefits. However, even the economic benefits were precarious; prestressed concrete always requires more costly equipment and far greater care to produce than ordinary reinforced concrete. In slender elements such as beams or slabs, for instance, even very slight variations in dimensions, in the positioning of the reinforcement, or in the quality or age of the concrete when removed from the formwork can lead to different degrees of camber being introduced when the prestressing is applied. This might not matter if the floor was to be concealed by a false ceiling, but in exposed structures such as multi-storey car parks (often built with precast, prestressed beams and floor units) careful selection of units of similar camber would be needed to avoid unsightly steps between adjacent units. Not all contractors were this conscientious.

Fig. 2.32 Roof/door detail on British Overseas Airways Corporation maintenance head-quarters hangar, Heathrow Airport, west London, 1950–1955. Engineer and architect Owen Williams (photo Bill Addis).

Fig. 2.33 Roof detail on BOAC maintenance headquarters hangar, Heathrow Airport, west London, 1950–1955. Engineer and architect Owen Williams (photo Bill Addis).

Fig. 2.34   Roof detail on British European Airways maintenance hangar, Heathrow Airport, west London, 1951. Engineer Alan Harris. Part demolished 1999–2000 (photo Bill Addis).

Fig. 2.35   Warehouse at Spekelands Road Freight Depot, Liverpool. Engineers A.J. and J.D. Harris. The inverted T-beams each comprise 11 precast concrete elements post-tensioned with four cables. They are 6 ft. (1.83 m) wide and vary in depth from 5 ft. 5 in. (1.65 m) at the haunches to 10 in. (0.25 m) at mid-span. The two-pin portal frame spans 102 ft. 6 in. (31.2 m) (photo Bill Addis).

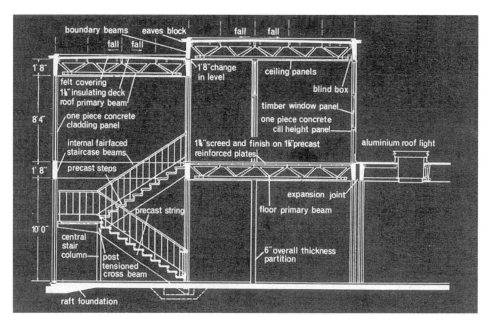

Fig. 2.36 Laingspan flooring and roofing system (from Deeson, 1964).

Bison Wet Cast Prestressed

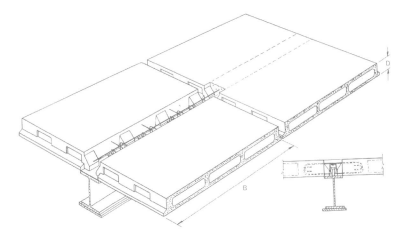

Fig. 2.37   A Bison pre-stressed concrete flooring system (from Deeson, 1964).

Concrete construction and design

Materials

In the 1950s and 1960s, the main developments in materials, and also in construction methods, were directed towards faster construction of re-inforced concrete structures. Various materials were employed to achieve this, the most notorious of which were high-alumina cement and the additive calcium chloride. Problems arising from their use did not become evident until the 1970s, and are discussed more fully in the section below.

Lightweight aggregate and lightweight concrete

The 1950s saw the introduction of a new lightweight aggregate, Lytag, developed from a waste product of coal-fired power stations (i.e. the dust extracted from flue gases). This was pulverised fuel ash (PFA), also known as fly ash. The ash is mixed with water, formed into small pellets and sintered (fired at a high temperature) to create lightweight pellets. This aggregate produced a concrete of comparable strength to concrete made with the normal coarse aggregate of crushed stone or gravel, but with a 30% lower density. This was an early and encouraging example of recycling an apparently unusable waste product. It also demonstrated that it was possible to make concrete without the need to quarry stone or dredge river gravels, processes that have such a dreadful impact on the natural environment.

Lightweight concrete has the great merit of reducing the structural weight and hence reducing the loads to be carried down through the structure and foundations into the ground. This reduces structural costs significantly. In addition, lightweight concrete has superior thermal insulation properties, which was a significant factor in its subsequent widespread adoption as the material of choice for the inner leaf of external walls in houses and other habitable building types.

Other materials were employed to produce varieties of lightweight aggregate, e.g. waste products such as blastfurnace slag, and natural materials such as expanded shale, perlite and vermiculite. These materials are converted for use as aggregates by heating, sometimes with the addition of water, to achieve irreversible expansion into a stable, lighter state (or 'allotrope').

Concrete could also be made lighter by using conventional coarse aggregate and simply omitting the fine aggregate (usually sand). This resulted in a voided concrete with adequate strength for use in walls, making it well-suited for use in housing. Originally conceived in The Netherlands, 'no-fines' concrete found some use in Britain in the 1920s, but its widest use occurred after the war. In particular, the building contractor Wimpey used the material for both low-rise and later high-rise housing, building several hundred thousand no-fines dwellings in all (Finnimore, 1989; Reeves and Martin, 1989; Williams and Ward, 1991).

Steel

After the Second World War, 'high tensile' steel became more readily available for use in reinforced concrete. The permissible stress for this was typically some 50% higher than for mild steel. The higher forces acting on the reinforcing bars required greater care during the making process to achieve an adequate bond and anchorage to the adjacent concrete. A traditional solution to this had been to bend the ends of the bars into an L- or U- hook, but this involved additional labour both in bending and in fixing within the formwork. Manufacturers added ribs or other profiles to the surface of hot-rolled bars, or alternatively cold-twisted round, square or chamfered square bars to give improved mechanical anchorage and reduce the need for hooks (Fig. 2.38).

Prestressed concrete

Higher-strength steel was also essential to the development of prestressed concrete, but two issues became apparent. The high-tensile steel was found to be prone to a modest 'creep', or relaxation under load, which

Fig. 2.38   Steel reinforcement used in the late twentieth century. Top, round mild steel; centre, low-alloy, high-yield steel; bottom, cold-worked, high-yield steel (British Cement Association).

led to some loss of prestressing. This problem was overcome by heat-treating the steel before use. Secondly, it was realised that the loss of pre-stress in concrete of ordinary strength, due to shrinkage and creep of the concrete, was large enough to render the prestressing of only modest benefit (Walley, 1996). This stimulated the development of better, higher-strength concretes that, on account of their higher cement contents and reduced water–cement ratios, also suffered lower creep and shrinkage losses than concrete of ordinary strength. Of no less importance was the development of more effective ways of tensioning the steel bars or cables, and anchoring these more effectively to maintain the prestressing. As with reinforcing systems at the turn of the century, this led to a spate of systems that competed for acceptance. Initially, these were imported from the Continent, but British systems soon became available (Walley, 1996; Andrew and Turner, 1985).

Ferrocement

Ferrocement is one of the few new cement-based materials that has been developed in the twentieth century. It was pioneered by Pier Luigi Nervi in the late 1940s and 1950s, initially to make concrete boats. It consists of a sand-based concrete with reinforcement of fine wire woven into a square mesh. Nervi's genius was to use it to make permanent, precast formwork for conventional concrete. It holds a very high-quality surface finish and is very strong and stiff for its weight. He used it for a series of spectacular roofs over exhibition and sports halls (Mainstone, 1998). Although none of his buildings were in Britain, his structures inspired many engineers and architects in the 1950s (de Maré, 1958). For a recent example of the use of ferrocement in England, see Figs. 2.53–2.55.

Construction methods

A large number of building systems has already been mentioned, and to this must be added techniques for speeding up the curing process to enable concrete floors to carry loads more quickly.

Prestressing techniques

Two methods of prestressing were used (Walley, 1996). In pre-tensioned work, the steel reinforcement usually consisted of bundles of small-diameter wires. This was placed in the formwork (usually stout, steel moulds), tensioned, and then anchored at the ends of the moulds before the concrete was placed. When the concrete had gained sufficient strength, the formwork sides and ends were released and the small-diameter wires were then bonded to the concrete. The tensioning forces were now balanced by compression stresses within the concrete. This method was favoured for factory-made precast floor units. Units of the same size could be produced with varying bending strength by varying the number and location of the wires. The most efficient layout involved long-line moulds, in which the concrete could be cut to length, like timber, to suit the purchaser's needs.

   One disadvantage of this approach was that the amount and location of the steel was usually constant throughout the length of the units. This is structurally inefficient, since the stresses in a beam or slab vary along the length. In ordinary reinforced concrete this is reflected in the layout of bars: the heaviest reinforcement is used at mid-span in a supported beam or at the root of a cantilever, where the bending forces are highest. Some precasting firms developed a 'deflected' prestressing system for beams, in which the pre-tensioned wires or strands could be altered in a

Fig. 2.39   Floor slab with sheaths for reinforcement in place (British Cement Association).

direction along the length of the beam by bearing against pins within the moulds. However, this generated large forces when the steel was pre-tensioned, necessitating very stout formwork.

The other method of prestressing, post-tensioning, offered a simpler technology, with the additional advantage of being usable for *in situ* concrete work (Fig. 2.39). In this process, the required profile of the pre-stressing steel was incorporated into the concrete section using oversize sheathing, typically thin-section, flexible, corrugated-steel tubing, to form a duct. When the concrete was of adequate strength, but still supported on falsework, prestressing steel was introduced into the duct. This could be stranded cables, wires or rods. The steel was then tensioned to the required level using jacks with calibrated gauges, and then anchored at the ends using wedges, nuts or another (usually patented) anchorage system (Fig. 2.40). The falsework could then be removed. Finally, the ducts were usually grouted, both to improve the durability of the steel and to provide supplementary anchorage to the steel in the (happily rare) event of anchorage failure or relaxation. Anchorages were usually also encased in concrete (Andrew and Turner, 1985).

Fig. 2.40   Post-tensioning and fixing cables in a long-span beam (British Cement Association).

Slipforming and lift slab construction

The move towards taller buildings encouraged the application of slipforming, a technique briefly described earlier, in which a vertical concrete structure is built in a quasi-continuous operation rather than in a series of discrete floor-by-floor stages. There are two basic methods. In one the formwork is advanced, almost continuously, by jacking against the concrete and its reinforcement so that the formwork climbs up the hardening concrete. In the other, the formwork is in at least two sections that are alternately 'leapfrogged' one above the other. Both methods offer greatly accelerated construction and eliminate the need for formwork and falsework support all the way down to ground level. However, they require the efficient organisation of labour and of materials delivery and placing: once you have started, particularly with the climbing method, you have to finish! Maintaining accuracy of line is also important, and care is needed when raising the formwork to ensure that the 'green' (still-hardening) concrete is not damaged by the movement.

Fig. 2.41 Slipformed concrete core at Barclays Bank Foreign Headquarters on the corner of Fenchurch and Gracechurch Streets in the City of London, under construction in 1968 (courtesy Ove Arup Partnership).

The early use of slipforming was for industrial structures such as grain silos and chimneys. The Post Office Tower was one of the more prominent buildings to have a slipformed core, an obvious application for a monolith such as this, but other buildings also benefited from this technique, which allowed the core structure of an office block (as it usually was) to be built up quickly (Fig. 2.41). This aided an early start on roof-level plant, including lifting equipment, which is always a lengthy task, while the office floors could be built up later around the core. Slipforming has recently returned to favour, and is used in conjunction with steelwork for the framing beams and columns.

In contrast, lift slab construction was patented as a method of making the floors of a multi-storey building at ground level, one above the other,

Fig. 2.42   Lift slab construction (British Cement Association).

and then jacking them up into position (Fig. 2.42). A typical vertical structure comprised steel columns and a slipformed core, to which the floors
were attached by brackets. Clearly this system imposed repetitive floor
layouts, but it eliminated falsework and allowed human activity to be
concentrated near ground-floor level, where it is safer and more easily
supervised.

Other formwork developments

Many attempts were made to improve the effective use of formwork on
the average concrete building constructed *in situ*. Too often, the formwork
was made of plywood and was damaged or destroyed as it was struck,
allowing it to be used only once. The carpenter played as large a role in

concrete construction as did the steelfixer and concretor, giving rise to the unkind, but not entirely inaccurate, remark that concrete structures were built twice, once in timber and then in concrete, and then the timber structure was thrown away.

Re-usable components developed for formwork and its supporting falsework included the ubiquitous 'Acrow' adjustable screw-prop, and aluminium beam systems. Table forms were prefabricated and were often in hinged sections for rapid erection and dismantling. 'Flying' forms were robust enough to be lifted around the site using a tower crane, the rapidly adopted 'beast of burden' on post-war building sites. This was also used to deliver bundles of reinforcement for fixing, and concrete skips loaded from the ready-mixed trucks that were now coming to serve nearly all but the largest or remotest construction projects.

Precast systems

The use of precasting in system building developed in two ways. In one, the components were delivered from a distant factory to the site, usually by lorry. This allowed efficient production, but sometimes led to delays and other problems when structural wall and floor units were not delivered in the right sequence. An alternative used by some contractors was to establish a 'site factory' where the bespoke unit types could be produced from robust moulds under the control of the site managers. The set-up costs for such a site factory dictated the minimum number of units required, which in turn indicated how many dwellings or tower blocks were needed for it to be economic. At least one well-known contractor reckoned that three 20-storey blocks of council housing were the minimum for cost efficiency, with the result that trios of towers rose in groups on the site of demolished Victorian terraced housing in many parts of London and some other cities. Steam-curing was often used on such sites to accelerate the production cycle.

Another area for the application of precast systems was in cladding (Morris, 1966, 1978). With Government encouragement, several building systems were developed by groups of local authorities for building schools and other projects. The best-known of these, CLASP (Consortium of Local Authorities Special Project), offered a kit of parts for the client and architect to 'pick and mix'. Precast concrete cladding panels were a popular part of the kit.

Foundations and groundworks

The support of larger, heavier buildings demanded improved foundations. In London, the prevalent London clay encouraged the development of the bored pile, which could be installed by equipment ranging from a small tripod rig to a tractor-mounted auger drill. Such piles may occupy such a large proportion of the footprint of an existing building that any new development may be constrained to re-use them and to accept the limitations on layout and building mass. However, the re-use of piles can reduce construction time and costs, and is an intelligent example of recycling.

The call for larger basements and the need to work close to existing buildings required new techniques to be developed in order to form large holes in the ground with complete safety. The diaphragm wall, using the ICOS system, was pioneered in Britain in 1961 for the Hyde Park Road underpass in London, and later became a common element of larger schemes. It involved digging a vertical trench, and replacing the exca-vated soil with liquid bentonite, a mineral with the property of being thixotropic (i.e. it behaves as a gel when at rest). Thus it held apart the sides of the trench which would otherwise have collapsed inwards if the trench were empty. When the trench had reached its full depth, a rein-forcement cage was lowered into place and then concrete was placed with a tremie pipe. This displaced the bentonite, which was collected, filtered to remove soil and re-used. This resulted in a strong vertical wall that would hold up the ground behind it as the basement was completed.

Concrete finishes

This period saw the beginnings of the subsequent rapid growth in the variety and use of exposed concrete finishes, on both precast cladding panels and *in situ* structures. The Cement and Concrete Association, funded by the cement manufacturers, played a major role here by offer-ing comprehensive advice to clients, architects, engineers and contractors. The range of options was remarkable. The variety of aggregates and cements available yielded a wide range of basic hues in addition to the customary grey: black, brown, yellow, white, green, and even pink and red were among the possibilities. Formwork of rough-sawn, planed or coated timber, or of steel, provided surfaces with textures ranging from the roughness of a sawn plank to the smoothness of glass.

Acid-washing and the use of retarders on the formwork surface (the latter delaying the setting of the surface layer of cement paste) permitted 'exposure' or featuring of the individual pieces of aggregate. This yielded a range of finishes from sandpaper fineness to the coarseness of the coarse aggregate, which in itself could have a variety of surface textures. Alter-natively, the use of bush- and pick-hammer tools on hardened surfaces chiselled away material, with effects varying from light random pitting to deeply exposed aggregate. Of course all of these techniques removed some of the concrete cover that provided corrosion and fire protection to the embedded reinforcement, so the competent designer would have to allow for this by specifying a greater thickness of initial cover.

Design

The first British Standard code of practice for the design of reinforced concrete, CP 114, appeared in 1948 (Bussell, 1996b). This dealt essentially with *in situ* work. It largely followed the pre-war codes, but incorporated guidance on water–cement ratio and on the cover appropriate to external or internally corrosive exposure conditions. It encouraged inspections every 3–5 years to identify cracking and corrosion of reinforcement, which by then was coming to be recognised as a growing problem. This enlightened guidance received an unenthusiastic response from the industry and building owners at large. Codes of practice for prestressed concrete (CP 115) and for precast concrete (CP 116) followed in 1959 and 1965, respectively.

The structural designer of the time was assisted by numerous books and design aids, ranging from individual tables and nomograms to larger manuals (Fig. 2.43). Notable amongst these are the very many authoritative books from Concrete Publications Ltd, some 85 according to a recent bibliography (Booth, 1998; Booth in Sutherland *et al.*, 2001). Many included relatively simple formulae for calculating forces, stresses and deflections in structures of unusual forms, such as domes and pyramid roofs. As such they were invaluable to the busy designer–detailer, who was still working with a slide-rule. Although the computer had arrived, it was still around the corner for most engineers even at the end of the period, and the electronic calculator was still in the future (early 1970s).

The recent past: from the mid-1960s to 2001

The use of concrete

Context

Technical and engineering developments in concrete construction during the last 30 years have been rather fewer and generally less easy to see in finished buildings than those from the tremendous period of progress during the 1930s and 1950s.

The aggressive marketing of building systems in the 1960s has died down, and been replaced by a greater concern for speeding-up the construction process and ensuring high-quality and, often, fair-faced (exposed) concrete. The greatest impacts have resulted from the improvements in construction plant and methods, and the management of on-site

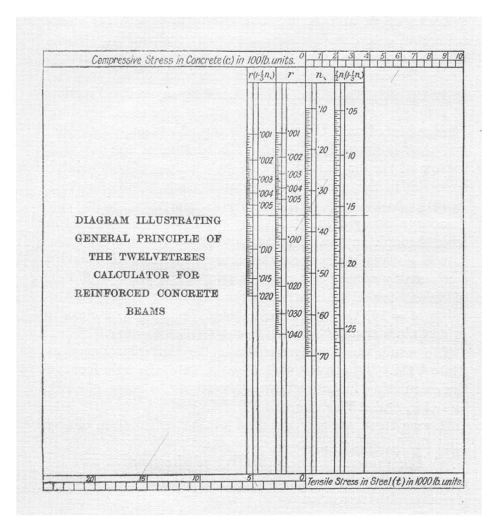

Fig. 2.43 Illustration of an early nomogram for design of reinforced concrete beams (from Twelvetrees, 1907).

building processes, often involving a great many different trades and sub-contractors (Illingworth, 2000).

The 1960s and 1970s saw an increasing acceptance of, and growing enthusiasm for, exposed concrete. This was true of both concrete cast *in situ*, often displaying on its surface the grain of the timber shuttering, and also an increasing variety of architectural cladding, essentially precast concrete panels attached to a concrete or steel frame (Morris, 1978; Ross, 1997; Stratton, 1997). During the 1970s, a number of buildings were adorned with a façade made from the newly developed glass-fibre-reinforced cement (GRC).

In the wake of the oil crisis in the early 1970s, when energy prices tripled, the most significant development in the use of concrete during the last decade or so has had little to do with its strength, its capacity for architectural and structural expression, or its potential for fast construction. It has been the thermal mass of reinforced concrete floors, which has enabled building services engineers to use them as an essential part of the heating and cooling system in so-called 'low-energy' buildings.

As if marking the centenary of reinforced concrete, however, the closing decades of the twentieth century have seen a dark cloud looming on the horizon. The long-term durability of concrete has become a major concern as many of the pre-war buildings and other structures have outlived their original purposes. It has become clear that the quality of pre-war concrete was far from consistent; some still looks as good as new, but rather more looks very much the worse for wear. This is usually caused by water penetration, the rusting of reinforcement and the consequent spalling of the surface concrete which accelerates the process of decay. Such damage can be a serious threat to a building's capacity to carry the loads imposed upon it, and it is unpleasant to look at.

Many post-war buildings have suffered in similar ways. A number of the building systems devised for rapid construction during the 1950s and 1960s have proved to be particularly vulnerable to water penetration, often leading to the rusting of the very steel that ensures the integrity of the whole building. Many such buildings have already been demolished, a reflection not only of their technical problems and the cost of repair and refurbishment, but also of the social failings of high- and medium-rise council flats.

The popularity of exposed concrete finishes in the 1970s and 1980s has also left its toll of unsightly buildings resulting from the inadequate handling of rainwater. This can cause concrete surfaces to become badly discoloured or stained, either by rain-borne chemicals or by impurities such as iron pyrites in the aggregates (Fig. 2.44).

Together, these problems with durability have led to concrete losing a lot of its former popularity. Steel has increased its share of the frame-building market and seriously eroded concrete's hold on the market for road and rail bridges. A successful marketing campaign by the steel industry has also encouraged the belief that steel is the structural material for fast construction. Concrete is suffering from a reputation for being a high-maintenance material which is difficult to make properly, and which restricts the flexibility and adaptability of buildings during a long life.

In fact, most of these generalisations about concrete are without foundation and are used, perhaps, to divert attention from a loss of site and project management skills. As the first users of concrete found, a hundred

Fig. 2.44   Surface staining in sheltered zone, not subject to regular rainwater washing (photo Susan Macdonald).

years ago, the quality of concrete building depends on the skill of the designers (engineer and architect), the site operatives and supervisors, and how well the concrete is maintained. The best examples today are no less outstanding than their ancestors from fifty or a hundred years ago.

Architecture

The 1960s was the decade when it finally became respectable to let concrete come out into the open. Although it found its most dramatic expression in massive, Brutalist buildings such as the Barbican and the National

Fig. 2.45   Langham Close, Ham, Surrey (1958) Architects James Stirling and James Gowan (photo Bill Addis).

Theatre in London, its beginnings were very restrained indeed. Three modest blocks of flats by James Stirling and James Gowan dared to reveal in their façade a 400-mm-deep concrete edge beam bearing the imprint of the timber shuttering (Fig. 2.45).

Within a few years, not only was structural concrete being put on show, but there quickly developed a new concrete cladding industry. This suited

Fig. 2.46 Glass-reinforced cement cladding on the Crédit Lyonnais building in Queen Victoria Street, City of London (courtesy Ove Arup & Partners; photo Mike Taylor).

the major contractors who were continuing to promote their various building systems which employed precast concrete elements. However, a great many failures occurred, caused mainly by corrosion of the steel fixings used to attach the concrete panels to the structural frame. There were also many successes, such as Crédit Lyonnais with its thin glass-reinforced cement (GRC) cladding panels, which have weathered well for more than 30 years (Fig. 2.46).

Amid these various developments in the visual aspects of concrete, this most recent period of concrete's history is most notable for the impact of the building services on the structure of buildings. These developments need special mention if only because the history of building services engineering is even less well charted and celebrated than that of structural engineering. Old services installations were usually removed at the end of their useful life, and built-in services such as ducts are inevitably less visible than structural elements.

Generally speaking, hardly any aspect of the original servicing of even a recent building will meet the demands of today's users.

First came the rationalisation of introducing modern services, especially air conditioning, into frame buildings. The multi-disciplinary firm of building designers Arup Associates deserves special mention. It was founded in 1963 by a group of structural and services engineers with the architect Philip Dowson, and operating as an entity which was separate from the consulting engineering practice. They sought to put into operation the philosophy of 'total architecture' that Arup himself had advocated, and set up Arup Associates to achieve this particular aim.

They began with a number of laboratory buildings for the expanding university sector, notably the Mining and Metallurgy Building at Birmingham University (Figs. 2.47 and 2.48), and these were followed by a number of buildings for accommodation and offices. The result was a series of highly acclaimed buildings which, step by step, developed more and more sophisticated ways of integrating all the engineering aspects of buildings to create a unified whole. Their technique for structuring the different zones within the building layout became known as the 'Tartan Grid' (Brawne, 1983).

Concrete was the ideal material with which to achieve this integration, since it can easily serve both load-bearing functions and the interpenetrating voids that are necessary to pass the services horizontally through the floor structure and vertically through risers that can usually also serve as shear walls for the building.

The conscious use of the building structure and fabric (especially concrete) to assist with the moderation of the internal environmental conditions of a building is often assumed to be a new idea. It is not: the Roman hypocaust is ample evidence of that. The early days of domestic concrete construction provide another example from 1870:

> The building is heated by air, warmed by contact with earthenware, conducted mainly through flues formed in the body of the concrete walls and admitted by sliding valvular gratings in the skirtings of several rooms. Ventilation is provided for in every room by distinct flues, formed in the concrete, and entered by apertures near the ceiling.

The conscious application of this philosophy in modern commercial buildings surfaced in the late 1960s. The Central Electricity Generating Board building by Arup Associates was among the first (commissioned 1973, completed 1978) (Figs. 2.49, 2.50).

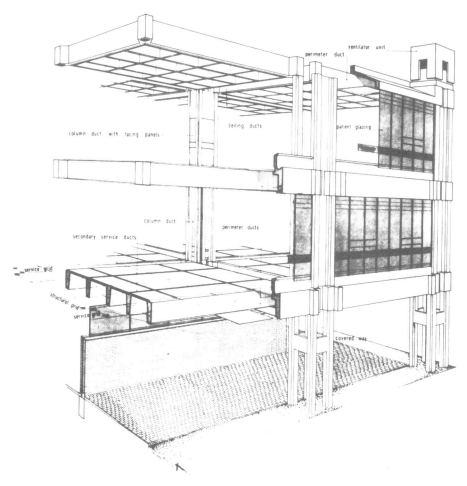

Fig. 2.47 Cut-away isometric of the Mining and Metallurgy Building, Birmingham University (photo Arup Associates).

However, these early pioneers were several years before their time. It is only since the mid-1990s that the significance of the life-long energy costs of buildings has come to be appreciated by (some) building owners and occupiers. First under the banner of 'green buildings', and more recently as 'sustainable buildings', architects and engineers can now offer clients buildings that consume less than a quarter of the energy of similar buildings only a decade ago.

In winter, the heat produced by people and computers in the building during the day is stored in the concrete structure and released to the air overnight, so the interior is already warm next morning. In summer, the concrete floors are cooled at night using outside air, so they are ready to

Fig. 2.48 Exterior view of the Mining and Metallurgy Building, Birmingham University (photo Arup Associates).

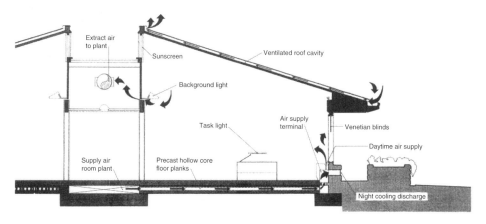

Fig. 2.49 Section of the CEGB building, Bedminster Down, Bristol (Arup Associates).

absorb heat produced during the following day and so reduce the need for expensive cooling.

This heat transfer is only efficient if the soffit of the concrete floor slab is exposed to the air, and the last few years have seen a number of build-

Fig. 2.50 Exterior of the CEGB building, Bedminster Down, Bristol (photo Arup Associates).

ings with beautifully finished and visually striking concrete ceilings. Just as concrete in the 1950s helped express building structure, so today it can help express the environmental qualities of a building (Figs. 2.51 and 2.52).

Structure

The most notable concrete structures of recent decades must surely be the concrete platforms developed to exploit the oil and gas resources under the North Sea, whose existence was virtually unproven at the start of the period under review. The scale of these structures dwarfs all but the tallest

Fig. 2.51 External view of the Wessex Water Headquarters building, near Bath, England. Architect Bennetts Associates; Engineers Buro Happold (courtesy Buro Happold; photo Mandy Reynolds).

Fig. 2.52 Internal view of the Wessex Water Headquarters building. Architect Bennetts Associates; Engineers Buro Happold (photo Mandy Reynolds).

buildings, while their design had to accommodate both their service as static platforms, under extreme climatic conditions, and their role as floating vessels while they were towed from construction basin to work site— the ultimate precast concrete unit! Much has been learned from the durability studies made prior to designing these structures, as well as the monitoring of their performance in service. This knowledge, together with a wealth of practical expertise, has been a rich source of new technology which has fed into the building industry.

Although concrete has been somewhat displaced by steel as the framing material for buildings in recent years, it must not be forgotten that it is still predominant in foundations, floor and roof slabs, and for many other applications. Even the recently rediscovered composite floor technology in steel-framed buildings combines light-gauge profiled metal decking with a far greater mass of concrete, and the commonest alternative flooring in steel-framed buildings is the precast concrete plank. Post-tensioned concrete floor slabs have also found some favour as an alternative to using long-span steel beams.

The characteristics of *in situ* flat slabs have often been exploited by architects. Their free-form potential was used to great advantage in the now Grade I listed Willis, Faber and Dumas building in Ipswich (1975, architect Foster Associates, engineer Anthony Hunt Associates), although the architect did not expose the soffit. The capacity of slabs to span and cantilever long distances was highlighted in the external galleries of the Royal National Theatre on London's South Bank (1976, architect Denys Lasdun and Partners, engineer Flint and Neill), where the exposed coffered slab soffit is an essential feature of this Brutalist landmark.

The advocates of concrete have made effective inroads by restating and re-presenting the potential of concrete in building. Funding of the Cement and Concrete Association was drastically cut in the mid-1980s, leading to the closure and sale of the highly regarded training centre near Slough. It has re-emerged in leaner, slimmer form as the British Cement Association. One topic in particular has been addressed by the BCA and others: fast construction in concrete (Bussell *et al.*, 1987). This has called for no new technology, but rather for a reminder of what is needed for fast building procurement whatever the materials involved, i.e. a simple, repetitive, buildable design.

Concrete construction and design

Materials

Research to improve the quality and mechanical properties of cements and concrete has continued, and the concretes used nowadays are much improved on their pre-war and 1960s ancestors (Somerville, 1996).

However, the increased use of concrete during the post-war period, and the pressure to build ever more quickly and cheaply, has brought its fair share of problems, which came to light following a number of failures during the 1970s.

High-alumina cement (HAC) has the useful property, when used in place of Portland cement, of producing concrete with a very high early strength. Accordingly, it was widely used in precast prestressed flooring manufacture, where it allowed the rapid removal of units from the form-work and a more productive use of the expensive forms. It was known, and recorded in codes of practice, that in warm, humid conditions HAC underwent so-called 'conversion', which resulted in a loss of strength.

The failure of several roof beams made with HAC in three structures in 1973–1974 led to more thorough investigations, which revealed that conversion was taking place in virtually all HAC structures over longer time periods. They also showed that conversion had not been the sole, or even the principal, factor in the failures; there was nevertheless wide-spread concern, as very many floors incorporated HAC units. Eventually it was accepted that the loss of strength resulting from conversion could be quantified.

Guidance was prepared that allowed most factory-made units to be declared safe (unless also subject to chemical attack, e.g. in particularly damp or aggressive environments). This guidance also offered advice on the appraisal of HAC structures generally (Department of the Environment and Welsh Office, 1975; Bate, 1984). The use of HAC in new structural work was effectively banned. To date, no further failures have been reported involving HAC.

Calcium chloride was another additive that was widely used to accel-erate the setting of cement, particularly in cold weather. It began to be widely used during the construction 'boom' from about 1960. Subsequent experience of the corrosion of reinforcement showed that this practice was unwise, and limits on its use were tightened. It was effectively banned in reinforced and prestressed concrete structures from 1977 (Pullar-Strecker, 1987).

Chlorides from other sources have caused severe damage to concrete structures. De-icing salt has seriously affected many bridges, poorly washed marine aggregates have carried salt into the concrete mix, and wind-blown salt spray has resulted in aggravated corrosion of exposed reinforced concrete, often some distance inland. Other material problems are discussed in Chapter 4.

It is no surprise that durability is now at the top of the agenda in rein-forced concrete design (see below). A further development in this respect has been the wider use of firstly galvanised steel, later stainless steel, and more recently epoxy-coated bars to provide robust protection against corrosion. Their higher cost has generally limited their use to locations where such protection is essential, such as road bridge decks and thin-section architectural cladding, although the application of life-cycle cost analysis can show that the higher capital outlay is merited by the reduction or elimination of expenditure on replacement or repair.

Modern cement is more finely ground than previously. This gives a more rapid development of strength, which is useful in allowing earlier striking of formwork, but there is a correspondingly reduced further increase in strength with time as the chemical reaction continues.

The durability requirements which favour increased cement content and a reduced water–cement ratio result, almost incidentally, in standard modern concrete typically having a cube strength as high as $30–35\,N/mm^2$, a far cry from the early days of the twentieth century when $11–15\,N/mm^2$ was the norm. Stronger concretes are routinely specified to achieve smaller column sizes, while research indicates that additives such as silica fume might soon make concrete with strengths of $80–100\,N/mm^2$ not uncommon.

Concrete blocks, and to a lesser extent concrete bricks, remain major elements of much domestic construction, other low-rise buildings, and partitioning. Used as alternatives to the fired clay brick, blocks, in particular, are made from different materials with a range of different densities, strengths and surface finishes. As such, they can perform numerous roles, for example in load-bearing walls, as facing skins with textured finishes, or as part of the thermal insulation of the fabric (Gage and Kirkbride, 1980).

A growing hostility to the environmental consequences of unlimited extraction of stone and gravel for aggregates has begun to direct attention towards the use of recycled aggregates, waste materials and other substitutes for newly won natural stone. Similarly, cement replacements using industrial waste products are both environmentally more responsible and more economical; happily, they sometimes bring technical benefits too, for instance by reducing the heat gain as freshly made concrete hydrates, thus reducing the risk of early-age thermal cracking.

Glass-reinforced cement (GRC) was developed to combine the compressive strength of cement and the tensile strength of glass fibres. The result is a mouldable material with very high strength and stiffness, suitable for use as cladding panels which can be finished with an extremely fine surface texture. The non-ferrous reinforcement is, of course, invulnerable to corrosion. Despite its potential, however, it is relatively expensive and, apart from its occasional use in façades since the 1970s (for example Crédit Lyonnais, see Fig. 2.46), it has found little enduring application in buildings.

Fig. 2.53 Exterior of the Schlumberger Research Headquarters, Cambridge (Architects Michael Hopkins & Partners; Engineers Buro Happold; photo Bill Addis).

The last few decades have also seen a few buildings making use of ferrocement, the material developed by Nervi in the late 1940s and 1950s. By forming it in steel or fibreglass moulds, a shell with a very high-quality surface finish is achieved, which provides a permanent formwork for the primary reinforced concrete which carries the main structural loads (Addis, 2001) (Figs. 2.53–2.55).

Overall, however, it has to be said that the major change in the industry during this period has been the progress towards achieving concrete with higher and more reliable quality and properties. This has come about by means of a number of revolutions in construction management, such as the adoption of quality assurance (QA) methods and certification procedures by which manufacturers and suppliers seek to ensure that their products are certified to comply with customers' specifications and regulations. Together, these have helped reduce the risk to clients of not achieving the required quality of concrete, or of a project overrunning its time and cost targets.

Construction methods

The need to achieve faster concrete construction, and some ways of achieving this, have been discussed above.

Fig. 2.54 Interior of the Schlumberger Research Headquarters, Cambridge (Architects Michael Hopkins & Partners; Engineers Buro Happold; photo Bill Addis).

Fig. 2.55 Ferrocement formwork for a floor slab of the Schlumberger Research Headquarters, Cambridge (Architects Michael Hopkins & Partners; Engineers Buro Happold; photo Buro Happold).

There has been a cautious return to the use of precast large-panel construction. An example is the 'tilt-up' method, in which the walls of a simple building such as a high warehouse can be precast on the ground outside its perimeter, and then rotated into their final locations and secured together (Fig. 2.56).

Precast concrete panels, now more commonly marketed as 'architectural cladding', have undergone something of a renaissance and several new companies have emerged. However, this period has also seen the demise of several old-established, highly respected cladding firms such as Empire Stone, Dean Jesmond and Shawells. Precast flooring units, stair flights and other repetitive components are in wide use.

For *in situ* work, the ready-mixed concrete suppliers have also embraced quality assurance (QA) as part of their aim to produce a consistent, reliable product. The pump has become a viable alternative to the crane-borne skip of concrete, allowing concrete to be delivered over long distances and at considerable height (Illingworth, 2000). Once in the formwork, the concrete must be spread, compacted, finished and cured properly. Attempts have been made to revive the craft status of what, since the First World War, has been seen as the unskilled trade of concretor. Several firms specialising in concreting work are again applying the specialist, high-quality approach so jealously guarded by Hennebique and his competitors in the early twentieth century (Fig. 2.57).

Fig. 2.56 'Tilt-up' construction (photo British Cement Association).

Fig. 2.57 St John's College, Oxford. Structural engineers Price and Myers; Architects Macormac Jamieson and Prichard; Concrete contractor Trent Concrete (photo Heini Schnebeli).

Design

A new 'unified' code of practice for concrete design, CP 110, appeared in 1972 (Bussell 1996b). This incorporated guidance on reinforced, prestressed and precast concrete, replacing CP 114, CP 115 and CP 116, respectively. The new code addressed robustness as an explicit requirement to ensure that structural elements are adequately tied together, as required by the recently amended national building regulations. CP 110 was superseded by BS 8110 in 1985, which for the first time gave priority to durability requirements; the designer was virtually obliged to select a minimum concrete strength to achieve the required durability before beginning to consider loads and stresses in the material. At the time of writing, this standard seems likely to be superseded in turn by the Eurocode for structural concrete, published as a so-called 'pre-standard' (DD ENV 1992–1) in 1992.

Three major changes in the early 1970s were significant for all concerned with concrete. Firstly, the metric system was gradually introduced into the building industry, to the vexation of those less flexibly minded or numerate. Secondly, CP 110 was a limit-state code, abandoning the permissible stress approach that had served since the dawn of reinforced concrete. In practice, this had little effect on designs, as the new code was 'calibrated' against its predecessors to ensure it gave similar results. However, matters such as the assessment of crack widths and deflections were more fully addressed, and required more calculation. This was, perhaps, tacit acknowledgement and exploitation of the third change taking place—the growing role of the computer as an aid to undertaking and producing calculations and to design in general. This process has advanced so rapidly that an engineer today can design, draw, calculate and reinforce a concrete structure touching only a keyboard and a mouse, rather than the pencil, slide-rule, drawing board and tee-square that were obligatory only two or three decades ago. The resulting design information can also be transmitted by e-mail in seconds, eliminating the printer, the clerk and the postman.

It remains to be seen whether these developments will affect the issues that face those investigating the pathology of today's concrete structures in the future.

References

Addis, W. (1997) Concrete and steel in twentieth-century construction: from experimentation to mainstream usage. In: Stratton, M. (ed), *Structure and Style*. Spon, London, pp. 103–142.

Addis. W. (2001) *Creativity and Innovation: The Structural Engineer's Contribution to Design*. Architectural Press, Oxford.

Allan, J. (1992) *Berthold Lubetkin*. RIBA Publications, London.

Anchor, R.D. (1996) Concrete shell roofs, 1945–1965. *Proceedings of the Institution of Civil Engineers: Structures and Buildings* **116**:381–389.

Andrew, A.E. and Turner, F.H. (1985) *Post-Tensioning Systems for Concrete in the UK: 1940–85, Report 106*. Construction Industry Research and Information Association, London.

Architects' Journal (1932) Messrs Boots Factory, Beeston, Nottingham. *Architects' Journal* **76**:125–136.

Architects' Journal (1938) Extensions, Boots Factory, Beeston, by Sir E. Owen Williams, KBE. *Architects' Journal* **88**:1053–1060.

Architects' Journal (1994) Factory with a facelift. *Architects' Journal* **200**:31–41.

Architects' Journal (1997) New formula for original chemistry. *Architects' Journal* **205**:32–35.

Arup, O.N. (1944) *Memorandum on Box Frame Construction for Terrace Houses and Flats*. Arup, London.

Bate, S.C.C. (1984) *High Alumina Cement Concrete in Existing Building Structures. Report BR235*. Her Majesty's Stationery Office, London.

Beatty, C.J.P. (ed) (1966) *The Architectural Notebook of Thomas Hardy*. Dorset Natural History and Archaeological Society, Dorchester.

Booth, L.G. (1998) Discussion on Bussell (1996b). The development of reinforced concrete: design theory and practice. *Proceedings of the Institution of Civil Engineers: Structures and Buildings* **128**:397–400.

Bowley, M. (1966) *The British Building Industry: Four Studies in Response and Resistance to Change.* Cambridge University Press, Cambridge.

Brawne, M. (1983) *Arup Associates: The Biography of an Architectural Practice.* Lund Humphries, London.

BS 15 (1906) *Standard Specification for Structural Steel for Bridges and General Building Construction.* Engineering Standards Committee, London.

BS 785 (1938) *Rolled Steel Bars and Hard-Drawn Steel Wire for Concrete Reinforcement.* British Standards Institution, London.

BS 882 (1940) *Coarse and Fine Aggregates from Natural Sources for Concrete.* British Standards Institution, London.

BS 1144 (1943) *Cold Twisted Steel Bars for Concrete Reinforcement.* British Standards Institution, London.

BS 1221 (1945) *Steel Fabric for Concrete Reinforcement (Covered Expanded Metal, as Well as Hard-Drawn Steel Wire and Twisted Steel Fabric).* British Standards Institution, London.

BS 8110 (1985) *Structural Use of Concrete.* British Standards Institution, London.

Brown, J.M. (1966) Wilkinson W.B. (1819–1902) and his place in the history of reinforced concrete. *Transactions of the Newcomen Society* **39**:129–142.

Building Industries National Council (1939) *Code of Practice for the Use of Reinforced Concrete in the Construction of Buildings.* Building Industries National Council, London.

Bussell, M.N. (1996a) The era of the proprietary reinforcing systems. *Proceedings of the Institution of Civil Engineers: Structures and Buildings* **116**:295–316.

Bussell, M.N. (1996b) The development of reinforced concrete: design theory and practice. *Proceedings of the Institution of Civil Engineers: Structures and Buildings* **116**:317–334.

Bussell, M., Fitzpatrick, A. and Whittle, R. (1987) Fast building in concrete. *Architects' Journal* **186**:55–63.

Chrimes, M.M. (1996a) The development of concrete bridges in the British Isles prior to 1940. *Proceedings of the Institution of Civil Engineers: Structures and Buildings* **116**:404–431.

Chrimes, M.M. (1996b) Concrete foundations and substructures: a historical review. *Proceedings of the Institution of Civil Engineers: Structures and Buildings* **116**:344–372.

Christophe, P. (1902) *Le béton armé et ses applications.* 2nd edn. Béranger, Paris.

Coignet, E. and Tédesco, N. de (1894) *Du calcul des ouvrages en ciment avec ossature métallique.* La Société des Ingénieurs Civils de France, Paris.

Collins, P. (1958) *Concrete: The Vision of a New Architecture.* Faber and Faber, London.

Committee on Reinforced Concrete (1910) *Interim Report on Reinforced Concrete.* Institution of Civil Engineers, London.

Cottam, D. (1986) *Sir Owen Williams 1890–1969.* Architectural Association, London.

CP 114 (1948) Codes of Practice Committee for Civil Engineering, Public Works and Building. *The Structural Use of Normal Reinforced Concrete in Buildings.* British Standards Institution, London.

CP 115 (1959) *The Structural Use of Prestressed Concrete in Buildings.* British Standards Institution, London.

CP 116 (1965) *The Structural Use of Precast Concrete.* British Standards Institution, London.

CP 110 (1972) *The Structural Use of Concrete.* British Standards Institution, London.

Crook, J.M. (1965) Sir Robert Smirke: a pioneer of concrete construction. *Transactions of the Newcomen Society* **38**:5–22.

Cusack, P. (1984) François Hennebique: the specialist organisation and the success of ferro-concrete: 1892–1909. *Transactions of the Newcomen Society* **56**:71–86.

DD ENV 1992-1: Eurocode 2 (1992) *Design of Concrete Structures.* British Standards Institution, London.

DD ENV 1992-1-1: Eurocode 2 (1992) *Design of Concrete Structures. General Rules for Buildings (Together with United Kingdom National Application Document).* British Standards Institution, London.

de Courcy, J.W. (1987) The emergence of reinforced concrete 1750–1910. *Structural Engineer,* **65A**, 315–322; Discussion (1988) *Structural Engineer* **66**:128–130.

Deeson, A.F.L. (ed) (1964) *The Comprehensive Industrialised Building Systems Annual 1965.* House Publications, London.

Delhumeau, G. (1992) Hennebique and building in reinforced concrete around 1900. *Rassegna* **49**(1):15–25.

de Maré, E. (ed) (1958) *New Ways of Building,* 3rd edn. Architectural Press, London.

Department of the Environment and the Welsh Office (1975) *Building Regulations Advisory Committee: Report by Sub-Committee P (High Alumina Cement Concrete).* Department of the Environment and the Welsh Office, London.

Diamant, R.M.E. (1964) *Industrialised building—50 international methods;* (1965) *Industrialised building (2nd series)—50 international methods;* (1968) *Industrialised building (3rd series)—70 international methods.* Iliffe Books, London.

Edgell, G.J. (1985) The remarkable structures of Paul Cottancin. *Structural Engineer* **63A**:201–207.

Engineering Standards Committee (1904) *British Standard Specification for Portland Cement.* Crosby, Lockwood & Son, London.

Finnimore, B. (1989) *Houses from the Factory: System Building and the Welfare State, 1942–74.* Rivers Oram Press, London.

Fisher Cassie, W. (1955) Early reinforced concrete in Newcastle upon Tyne. *Structural Engineer* **33**:134–137.

Francis, A.J. (1977) *The Cement Industry 1796–1914: A History.* David & Charles, Newton Abbot.

Gage, M. and Kirkbride, T. (1980) *Design in Blockwork,* 3rd edn. Architectural Press, London.

Godwin, G. (1836) Prize essay upon the nature and properties of concrete, and its application to construction up to the present period. *Transactions of the Institute of British Architects* **1**:1–37.

Gray, A.S. (1988) *Edwardian Architecture: A Biographical Dictionary,* 2nd edn. Wordsworth, Ware.

Griffiths, H., Pugsley, A. and Saunders, O. (1968) *Report of the Inquiry into the Collapse of Flats at Ronan Point, Canning Town*. Her Majesty's Stationery Office, London.

Guillerme, A. (1986) From lime to cement: the industrial revolution in French civil engineering (1770–1850). *History and Technology* **3**:25–85.

Hamilton, S.B. (1956) *A Note on the History of Reinforced Concrete in Buildings: National Building Studies Special Report No. 24*. Her Majesty's Stationery Office, London.

Higgs, M. (1960) Felix James Samuely. *Architectural Association Journal* **76**:2–31.

HMSO (1909) *London County Council (General Powers Act)*. His Majesty's Stationery Office.

Hurst, B.L. (1996) Concrete and the structural use of cements in England before 1890. *Proceedings of the Institution of Civil Engineers: Structures and Buildings* **116**:283–294.

Hurst, L. (1998) Edwin O. Sachs—engineer and fireman. In: Wilmore, D. (ed) *Edwin O. Sachs: Architect, Stagehand, Engineer & Fireman*. Theatresearch, Summerbridge, pp. 120–131.

Hyatt, T. (1877) *An Account of Some Experiments with Portland-Cement Concrete, Combined with Iron as a Building Material, with Reference to Economy of Metal in Construction, and for Security Against Fire in the Making of Roofs, Floors and Walking Surfaces*. Chiswick Press, London (reprinted by American Concrete Institute, Detroit, 1976).

Illingworth, J. (2000) *Construction Methods and Planning*. 2nd edn. Spon, London.

Joint Committee on Reinforced Concrete (1907) Report of the Joint Committee on Reinforced Concrete. *Journal of the Royal Institute of British Architects* (3rd ser.) **14**:513–541 (discussion 497–505).

Joint Committee on Reinforced Concrete (1911) *Second Report of the Joint Committee on Reinforced Concrete*. Royal Institute of British Architects, London.

Jones, B.E. (ed) (1920) *Cassell's Reinforced Concrete*, 2nd edn. Waverley, London.

London County Council (1915) *Reinforced Concrete Regulations*. London County Council, London.

Mainstone, R.J. (1998) *Developments in Structural Form*, 2nd edn. Architectural Press, Oxford.

Mallory, K. and Ottar, A. (1973) *Architecture of Aggression: A History of Military Architecture in North West Europe 1900–1945*. Architectural Press, London.

Marsh, C.F. (1904) *Reinforced Concrete*. Constable, London.

Marsh, C.F. and Dunn, W. (1908) *Manual of Reinforced Concrete and Concrete Block Construction*. Constable, London.

Morice, P.B. and Tottenham, H. (1996) The early development of reinforced concrete shells. *Proceedings of the Institution of Civil Engineers: Structures and Buildings* **116**:373–380.

Morris, A.E.J. (1966) *Precast Concrete Cladding*. Fountain Press, London.

Morris, A.E.J. (1978) *Precast Concrete in Architecture*. George Godwin, London.

Mouchel, L.G. and Partners (1917) *Standard Specification for Ferro-Concrete*. L.G. Mouchel & Partners, London.

Newby, F. (1996) The innovative use of concrete by engineers and architects.

Proceedings of the Institution of Civil Engineers: Structures and Buildings **116**:264–282.

Payne, A. (1875) Concrete as a building material. *Transactions of the Institute of British Architects*, pp. 179–192; 225–254.

Perry, V. (1994) *Built for a Better Future: The Brynmawr Rubber Factory*. White Cockade, Oxford.

Port, M.H. (1995) *Imperial London: Civil Government Building in London 1851–1915*. Yale University Press, London.

Powell, C. (1996) *The British Building Industry Since 1800: An Economic History*, 2nd edn. Spon, London.

Pullar-Strecker, P. (1987) *Corrosion Damaged Concrete: Assessment and Repair*. CIRIA/Butterworth, London.

Reeves, B.R. and Martin, G.R. (1989) *The Structural Condition of Wimpey No-Fines Low-Rise Dwellings. Report BR 153*. Building Research Establishment, Garston.

Reinforced Concrete Structures Committee (1933) *Report of the Reinforced Concrete Structures Committee of the Building Research Board*. His Majesty's Stationery Office, London.

Ross, P. (1997) The relationship between building structure and architectural expression: implications for conservation and refurbishment. In: Stratton, M. (ed) *Structure and Style*. Spon, London, pp. 143–163.

Saint, A. (1976) *Richard Norman Shaw*. Yale University Press, London.

Sharp, B.N. (1996) Reinforced and prestressed concrete in maritime structures. *Proceedings of the Institution of Civil Engineers: Structures and Buildings* **116**:449–469.

Simmonet, C. (1992a) Le béton Coignet. *Les Cahiers de la Recherche Architecturale* **29**:15–32.

Simmonet, C. (1992b) The origins of reinforced concrete. *Rassegna* **49**(1):6–14.

Skempton, A.W. (1966) Portland cements 1843–1887. *Transactions of the Newcomen Society* **35**:117–152.

Somerville, G. (1996) Cement and concrete as materials: changes in properties, production and performance. *Proceedings of the Institution of Civil Engineers: Structures and Buildings* **116**:335–343.

Stanley, C.C. (1979) *Highlights in the History of Concrete*. Cement and Concrete Association, Slough.

Stratton, M. (ed) (1997) *Structure and Style*. Spon, London.

Sutherland, R.J.M. *et al.* (2001) In: Sutherland, J., Humm, D. and Chrimes, M. (eds) *Historic Concrete: Background to Appraisal*. Thomas Telford, London.

Twelvetrees, W.N. (1907) *Concrete–Steel Buildings*. Whittaker, London.

Walley, F. (1996) Prestressing. *Proceedings of the Institution of Civil Engineers: Structures and Buildings* **116**:390–403.

Wayss, G.A. (1887) *Das System Monier*. Wayss, Berlin.

Williams, A.W. and Ward, G.C. (1991) *The Renovation of No-Fines Housing. Report BR 191*. Building Research Establishment, Garston.

Witten, A. (1996) The Concrete Institute 1908–1923, precursor of the Institution of

Structural Engineers. *Proceedings of the Institution of Civil Engineers: Structures and Buildings* **116**:470–480.

Further reading

Arup, O.N. (1939) *Design, Cost, Construction and Relative Safety of Trench, Surface, Bomb-Proof and Other Air-Raid Shelters.* Concrete Publications, London.

Collins, A.R. (ed) (1983) *Structural Engineering—Two Centuries of British Achievement.* Tarot Print, Chislehurst.

Cusack, P. (1987) Agents of change: Hennebique, Mouchel and ferro-concrete in Britain 1897–1908. *Construction History* **3**:61–74.

Davey, N. (1961) *A History of Building Materials.* Phoenix House, London.

Faber, O. (1949) The structural use of normal reinforced concrete in buildings. *Structural Engineer* **27**:193–208.

Morreau, P. (ed) (no date) *Ove Arup 1895–1988.* Institution of Civil Engineers, London.

Newby, F. (ed) (2001) *Early Reinforced Concrete: Studies in the History of Civil Engineering, Vol. 11.* Ashgate Variorum, Aldershot.

Reynolds, C.E. (1932) *Reinforced Concrete Designer's Handbook.* Concrete Publications, London (later editions up to 1988).

Stratton, M. (1997) Clad is bad? The relationship between structural and ceramic facing materials. In: Stratton, M. (ed) *Structure and Style.* Spon, London, pp. 164–192.

Strike, J. (1991) *Construction into Design: The Influence of New Methods of Construction on Architectural Design 1690–1990.* Butterworth–Heinemann, Oxford.

Walley, F. (1962) *The Progress of Prestressed Concrete in the United Kingdom.* Cement and Concrete Association (Prestressed Concrete Development Group), Slough.

White, R.B. (1965) *Prefabrication: A History of its Development in Great Britain. National Building Studies Special Report No. 36.* Her Majesty's Stationery Office, London.

Many architectural and civil engineering periodicals are rich in articles dealing both with the general practice of concrete design and construction, and with specific buildings and structures. Those of particular use are listed below, in chronological order of first appearance either of the named periodical or of a predecessor of that periodical.

Proceedings of the Institution of Civil Engineers (1837–present).

Building (1842–present; originally *The Builder*).

The Architects' Journal (1895–present).

Concrete and Constructional Engineering (1908–1966).

The Structural Engineer (1921–present; journal of the Institution of Structural Engineers).

Part Two

The Appraisal, Assessment and Repair of Concrete Buildings

Chapter 3

Structural Appraisal

Michael Bussell

Introduction

Structural appraisal is an essential task in the repair and remediation of any existing concrete building or structure. The term describes the process of assessing the actual condition of a structure in relation to its use (whether an existing use, or a proposed change of use) in order to determine whether the structure can sustain that use. Whatever the reason for the appraisal, it involves the building pathology approach to determine structural adequacy. As such, the finding from the appraisal may be that the structure is satisfactory and requires no intervention in order to fulfil its future use. Alternatively, repair, strengthening or alteration may be necessary. As these actions will cost money and take time, it is clear that the appraisal has a pivotal role in deciding the structure's future—or indeed in deciding whether it has one.

This chapter outlines the circumstances in which a structural appraisal of a concrete structure might be made, and then describes the planning and conduct of the various stages of a typical appraisal. Not all stages will be necessary or appropriate in every case. Appraisal issues applicable to all buildings, such as investigation and assessment of ground conditions, are not considered here, and general survey and assessment techniques are discussed in the book *Building Pathology: Principles and Practice* (Watt, 1999). However, since virtually all twentieth-century structures, and many earlier ones, have concrete foundations of one type or another (irrespective of the construction materials used above ground level), these are considered below.

A fuller treatment of the general approach to structural appraisal is provided by the guide *The Appraisal of Existing Structures*, published by the Institution of Structural Engineers. The first edition of this guide was published in 1980, and was prepared by a Task Group under the chairmanship of the late Prof. Sir Edmund Happold. A revised edition, which appeared in 1996, incorporates much new guidance, and benefits from feedback and experience in using the first edition (Institution of Structural Engineers, 1996).

The need for an appraisal

An appraisal may be necessary for one of a number of reasons, which can broadly be grouped under two headings. One is establishing the construction, condition and behaviour of the structure as existing, with a view to continued use unchanged. The second is establishing the *status quo* as a preliminary to a change in use or other alteration to the existing structure. Building pathology is at the core of the task in all cases.

One of a number of 'interested parties' might initiate the appraisal. Typical sources for the work are noted for each of the likely reasons for an appraisal listed below. In practice, the appraisal appointment and brief would often come from an agent acting for the relevant party, such as a solicitor or surveyor.

Reasons for appraising the structure as existing

To provide a current record of construction and adequacy

A typical example of this is at the beginning of a lease in order to have a 'datum' record to be referred to when preparing a schedule of dilapidations at the end of the lease period. This is usually initiated by the current or intending owner or tenant, and is best conducted jointly for both parties, so that the record can be agreed as it is being compiled. (This is preferable to, and usually cheaper and quicker than, the alternative of separate inspections on behalf of both parties, especially when the inspections involve external access at high levels, or other hazardous and disruptive work. Separate inspections almost inevitably give rise to disagreement, followed by discussion(s), and eventually—it is to be hoped!—a resolution of differences of opinion.) Such work will rarely involve a detailed assessment of structural adequacy.

To provide advice on the structure

This might inform a decision regarding its purchase, leasing, mortgaging, insuring and the like. This is usually initiated by a prospective purchaser, tenant, finance house or insurance company.

To check the structure for adequacy

This is usually initiated by the owner or tenant following some damaging event such as a fire (Concrete Society, 1990), explosion, vehicle impact, or evidence of distress or deterioration. Some repair or remedial work might be needed as a result, the essence being to return the structure to its previous satisfactory state.

To check the structure for the presence of a material or a construction technique

This may be required following direction or advice that particular materials or construction methods may have deleterious effects on structural adequacy or durability. Examples from recent decades include the use of high alumina cement and alkali–silica reactions in concrete (both discussed in Chapter 4), and large-panel system-built high-rise housing blocks following the Ronan Point collapse (see Chapter 2). Direction on such cases has often been made by the relevant Government department to local authorities or other publicly funded bodies, with advice being concurrently offered to private sector owners or tenants.

Reasons for appraising the structure as existing, as a preliminary to a change in use or other alteration

To consider the effects of change of use

National building regulations (such as in England and Wales) do not, in general, demand the up-grading of existing structures to comply with subsequent changes in technical requirements. However, for certain 'material' changes of use (conversion to hotel, institution or public building, for example) it is mandatory to demonstrate compliance with the relevant current structural requirements. In such cases, the appraisal is usually initiated by whoever is responsible for the proposed change of use, occasionally following a 'prompt' from the building control authority if the appraisal has been overlooked! Structural alterations will generally be involved, in particular structural fire-protection measures.

To consider the effects of increased loading

This calls for a structural appraisal even when use does not change, as, for example, in a 1950s reinforced concrete office block now required to house modern automated document storage and retrieval units producing high local floor loading. This is usually initiated by the occupier seeking the change—owner or tenant, as applicable.

To consider the effects of alterations

This again will call for a structural appraisal even where use does not change. Careful thought must be given to any wider consequences of a local change. A case in point is the forming of an opening in one bay of a reinforced concrete floor slab, such as for a new stair. If the slab bay adjoins a cantilevered slab section, such as a balcony, the removal of concrete to form the opening might reduce the restraining effect of the bay on the cantilevered area, with potentially serious consequences of alarming sagging, or even collapse. The appraisal is usually initiated by whoever (owner or tenant) is proposing the alterations.

Developing the brief

Whoever commissions the work giving rise to the structural appraisal (the client) should be made aware of the need for an appraisal. The organisation occupying buildings or involved in their ownership or development may have professional advisers, either in-house (such as building or estates manager) or external retained consultants. These should be able to identify the need for an appraisal, and develop a brief for it. A 'lay' client should already have engaged an architect or surveyor who can advise likewise.

The brief for the appraisal must be clear in its aims and scope, and should give the reason(s) for the appraisal's being undertaken. It is also essential to give an adequate description of the structure(s) to be appraised. Nowadays, it is customary for such work to be put out to tender, often competitively, and so the bidders should be given as much information as possible to allow the submission of a realistic price and programme. Wherever possible, bidders should be given the opportunity to visit and inspect the structure so that they can assess for themselves how the work is to be done, and reflect this in a realistic bid price.

The form of appraisal report required should also be defined in terms of what must be included, and how many copies are required. Those experienced in appraisal will normally decide when bidding what will be provided in terms of illustrations (photographs, drawings, and so on) as well as how much of the detailed records made during the appraisal will be included, if this is not prescribed. Nowadays, it is not uncommon for an informed client to require an electronic version of the entire report on disk to allow further copies to be run off in the future if needed. (It should be remembered, however, that the copyright of the report remains with the originator, unless specifically re-assigned by agreement.)

A particular issue over which confusion can arise is that of a measured record of the building or structure. The client may assume that this will be produced as a matter of course, while those experienced in appraisal will be only too aware that this is a time-consuming, labour-intensive and hence expensive task that would be undertaken only if explicitly required, and hence priced for.

It is strongly urged that the term 'structural survey' is not used in describing such work, as both the phrase, and indeed both words on their own, have an imprecise or rather an 'elastic' meaning.

This is evident from the case of the lay house-buyer, who might believe that the brief valuation inspection of a property made by a building society surveyor obviates the need for a more thorough inspection made by a surveyor under instructions from the buyer. Neither of these inspections—commonly known as 'surveys'—can be equated with the more comprehensive work usually involved in an appraisal. Even if the appraisal is limited to a visual examination, the term 'inspection' is more precise and should be used in preference to 'survey'.

Similarly, the word 'structural' has a different meaning for the civil/ structural engineer and the lay person. The former normally takes it to refer to the elements of the building essential for it to stand up and be stable: floor and roof slabs or boarding and joists, beams, columns, trusses, loadbearing walls, foundations and so on. The lay person might assume that it also embraces floor, wall and ceiling finishes, partitions, glazing, joinery and so on. The civil/structural engineer will argue that these are not 'structural' as defined above, and are therefore outside the brief (and, indeed, often outside the engineer's field of competence, being more a matter for an architect). Again, it is essential to be precise in the requirements of the brief.

Access, and limitations on access, should be made clear. For example, it is not uncommon for a prospective purchaser or tenant to commission an inspection despite the present owner or occupier forbidding the removal of any finishes, including the lifting of carpets or taking-down of ceiling tiles. The findings of such an inspection will therefore necessarily be limited. Conversely, if a full inspection is required of the concrete cladding panels of a high-rise office block on a cramped site at a busy city road junction, then the time-scale and cost of the work should reflect the need for rope-access inspection—probably the only practical approach in this instance. Such an inspection is relatively slow, and can be disrupted by high winds or poor weather. The alternatives for access would be scaffolding, or a 'cherry-picker' or mobile access platform. The former would be inordinately expensive, while the latter might be ruled out on grounds of traffic disruption.

Bearing this in mind, the typical issues that must be clearly defined include those listed below, and these will also influence the level of information and advice that the appraisal will generate.

- What access will be permitted internally (e.g. all spaces, typical rooms only, removal of finishes for inspection only, ditto with permission to take samples for testing)?
- What means of access will be permitted internally (steps, ladder, platform, scaffolding, especially for high spaces)?
- What access will be permitted externally (e.g. all facades, selected 'easy-access' locations only, removal of finishes for inspection only, ditto with permission to take samples for testing)?
- What means of access will be permitted externally (steps, ladder, platform, scaffolding, rope access, mobile access platform)?
- Who is to provide the means of access?
- Who is to pay for providing the means of access?
- Who is to make good areas where sampling or other work has taken place?
- Who is to pay for this?

Where the brief is not well drawn, the conscientious consultant should raise, or seek clarification of, the issues that need to be defined. Less-conscientious folk may seek to exploit the situation by imposing extra charges afterwards on the grounds—possibly true—that extra work was needed, which was not defined in the brief.

Generally, health and safety regulations must be obeyed, particularly in regard to access. The scale of an inspection may well be large enough to bring it within the ambit of the Construction (Design and Management) Regulations 1994 (Health and Safety Commission, 2001), but even when these regulations do not formally apply, their philosophy is sound and offers useful guidance for the smallest investigation.

Commissioning suitable consultants

It is important to choose suitable people to undertake the appraisal; that is, they should have the requisite expertise and experience to carry out the work. A capable architect or surveyor can undertake member sizing of routine timber and masonry structures, aided by tables and other guidance in building regulations that are based on past experience and sound 'rules of thumb'. In contrast, reinforced concrete structures are, like structural steelwork, invariably 'engineered'. That is to say, they are sized and designed to carry specific loads with an adequate but economic margin of safety. There are no tables for this in building regulations, nor are there 'rules of thumb' to be applied by those not having a reasonable understanding of reinforced concrete design.

A further consideration with concrete structures is that a concrete element is 'opaque', i.e. its reinforcement is not visible, and simply measuring the cross-sectional dimensions cannot give an indication of its strength. The same problem applies to concrete-encased steel sections, and to a much lesser extent to cavity masonry walling. Whereas timber elements, solid masonry construction, and uncased steel or iron sections can essentially be defined by their overall size, reinforced concrete and cased steelwork must be opened up to establish the components that contribute so much to element strength. The problem is aggravated for reinforced concrete by the fact that reinforcement is not constant throughout most elements (particularly beams and slabs), being varied to match the applied forces in the interests of economy. In addition, runs of reinforcement are usually 'lapped' at bar ends, i.e. overlapped to provide continuity of robustness. Opening-up at a lap position might persuade the inexperienced investigator that there is twice as much reinforcement in the typical section than exists in reality.

All of this points to the need for the involvement of a suitably experienced civil or structural engineer. The services of such folk were, until fairly recently, to be found almost exclusively in one of the many consulting engineering practices, but today these have no monopoly. The multi-disciplinary building design firm has become more common, while building surveyors and the infant profession of 'facilities managers' also often has in-house civil and/or structural engineering capability.

Whoever is to be appointed to carry out the appraisal, what matters is that the individual(s) undertaking it should be knowledgeable, and competent to do such work. Evidence of this should be sought by the client, for example by a CV recording relevant expertise and previous experience.

Planning the appraisal

In the planning of the appraisal process, the work should be considered in its entirety, including all the necessary investigations (physical and documentary) needed to obtain the information required for assessment, followed by consideration of repair and/or other work necessary.

It is stressed that not all stages and activities described here may be necessary in each particular case. The scope of the appraisal will obviously depend on its specific objective(s). Equally, the appraisal may be limited to only a part of the structure. Each appraisal must be planned, and its scope determined, taking account of the individual circumstances. In particular, an 'inspection' might be limited to a walk-round.

It is recommended that appraisal work should generally be undertaken in the following order:

- initial inspection and appraisal;
- urgent action;
- information gathering, research and documentary review;
- detailed investigation;
- assessment of structural condition;
- consideration of work needed;
- reporting and recommendations.

The recommendations may include, in particular:

- further investigations, including load testing.

Experience has shown that this sequence offers an efficient working pattern. It is incorporated in *Building Research Establishment Digest 366: Structural Appraisal of Existing Buildings for Change of Use* (BRE, 1991), which is cited as guidance on dealing with existing buildings in *Approved Document A: Structure* of the current building regulations (Department of the Environment and Welsh Office, 1994).

The description of a typical appraisal sequence offered below is based on looking at a single building or structure. Thought must be given to the conduct of the work when dealing with larger structures such as a 20-storey block of system-built flats, or when there are many structures of similar construction, such as a local authority housing estate. In such a case it is generally appropriate to focus in detail on several 'case studies' of visibly varied condition, and then to deal with the remaining units by less extensive (but not casual!) inspection, with more detailed work being reserved for any additional problems or anomalous construction observed. A standardised inspection sheet, prepared after the first few cases have been studied, will be a valuable aid to effective working. Nowadays, this might well be completed in electronic form on a hand-held computer as the inspection is carried out.

It is worthwhile to make an early approach to the building control authority if the results of the appraisal are to be submitted for regulatory approval. This will allow any particular requirements for testing, and so on, to be incorporated in the appraisal.

Likewise, if a listed building, or a building within a conservation area, is being investigated, then it is prudent to discuss the works with the local

Fig. 3.1 Falmer House, University of Sussex, designed by Sir Basil Spence 1960–1961, is a listed Grade 1 building. Testing can be an invasive process, and when dealing with a listed building special care must be taken to design the testing so that it minimises the effect on the building. In some cases, statutory consent may be required (photo Susan Macdonald).

authority and/or national heritage body (such as English Heritage or Historic Scotland). Listed building consent is needed for any alterations to a listed building that affect its character as a building of architectural or historic interest. Rigorously, 'alterations' can be understood to include the removal of even small samples for testing, so it is advisable to obtain consent to such work. Planning permission may be required if any changes are proposed to the appearance of a building that is in a conservation area (Fig. 3.1).

Initial inspection and appraisal

An initial inspection is always valuable. It will give a 'feel' for the structure and its condition, and assist the detailed organisation of the appraisal. Notes, sketches and photographs will be valuable aides-mémoire. Problems such as distortion, damage, rusting and so on may be evident, again informing the planning of the work. In cases where the structure has suffered damage from an event such as a fire, explosion or vehicle impact, then it will be an obvious priority to examine the affected area(s) and to consider whether strength or stability have been compromised. If so, urgent action will be required, as discussed below.

This initial inspection is also a good opportunity to become acquainted with those using the building, and to explain what will follow in terms of subsequent site work. Good relations established now will smooth the future path. Often, too, such conversations will reveal information about the structure that the instructing client is either unaware of, or chooses not to disclose! (For example, persistent leakage which could be causing corrosion of reinforcement, repeatedly jamming doors or windows that might suggest continuing foundation movement, or 'the oldest inhabitant' remembering alterations or repair work carried out in the distant past and otherwise unrecorded.)

It is invariably useful to consult records already available (drawings, past reports and so on). Not least, this allows a cross-check between drawings and the visible reality, which will show whether the drawings are up-to-date and otherwise reliable. Drawings and other records held on the premises should be sought. Again, these may not be known to the instructing client.

If no records are available at this stage, it will be useful to look critically at the structure in order to estimate its probable construction date and to identify any obvious construction forms and techniques. Examples of the latter are precast concrete cladding panels, and shallow long-span floor or roof beams, suggestive of prestressing. Such information will be valuable when it comes to gathering information on the structure (see below). The experienced appraiser will also be aware of potential concrete problems associated with the particular construction period, as discussed in Chapters 2 and 4.

There is usually minimal scope for making calculations on a reinforced concrete structure at this stage unless original construction drawings, including reinforcement details, are to hand. However, the experienced eye should be looking for matters to be noted and reported. An example is the 'droopy' cantilever, a consequence of long-term deflections due to concrete shrinkage and creep (see Chapter 4). Balconies in structurally adventurous blocks of flats are possible sufferers from such 'droop', which is not usually a concern on grounds of safety or strength, but will need to be pointed out to a prospective purchaser or lessee.

The client might well ask for an instant opinion following this initial inspection. Unless there is an evident problem demanding urgent action (see below), it is preferable to refrain from anything but a very general view, suitably qualified. Suitable wording might be along the lines of 'I haven't seen any obvious major problems at this time, but I will of course need to investigate further before I can give you a fuller report'.

Urgent action

Sometimes the initial inspection will reveal a potentially or actually dangerous state of affairs, requiring immediate action. This might certainly be the case when inspecting a building after fire, explosion or vehicle impact. Action in such an event could include immediate evacuation and/or emergency safeguarding works such as propping or shoring, and protective fencing-off of the dangerous structure at a sufficient distance to ensure public safety if the worst should happen and total collapse should occur.

More commonly, the initial inspection might reveal potential hazards such as cracked and loose lumps of concrete spalled off by reinforcement corrosion, which could fall and cause injury, or worse, to people, and damage to property. As an example, the concrete cladding of a large seafront building, somewhat in need of maintenance, was inspected from ground level using binoculars. This indicated numerous pieces of loose concrete, corrosion of the steel having been aggravated by salt spray. A 'cherry-picker' was immediately hired to give access to all elevations, and the loose lumps of concrete were removed by hand. The resulting pile of concrete weighed over half a ton.

Information gathering, research and documentary review

As already noted, reinforced concrete is an 'opaque' material in which the pattern of embedded reinforcement is not visible. Non-destructive testing techniques can be helpful in establishing this pattern, with opening-up on a spot-check basis to confirm the findings. However, it will invariably be worth seeking original drawings of the structure and other relevant documentary information, and effort put in this direction will be well rewarded.

There are two sorts of information to be considered. These are building-specific information, and generic information relating to the design and construction practice of the time.

Building-specific information

Relevant building-specific information includes:

- drawings (especially of the concrete structure and its reinforcement);
- structural calculations;
- reinforcement bending schedules;
- structural specifications;

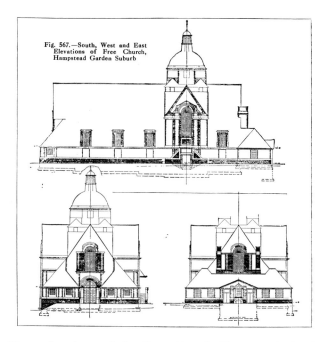

Fig. 3.2 These drawings of the Free Church Hampstead Garden Suburb, designed by E. Lutyens, were published in 1920. There is also a photograph of the building under construction. Contemporary books and journals can be an important source of information on individual buildings (from Jones 1920).

- reports and other records of inspections, maintenance, repairs, alterations, etc.;
- published accounts of the structure in technical journals.

Where is this information likely to be found? The obvious answer is that it will (or should) be with the designer of the structure. It is important, therefore, to try to identify who this was; not always an easy task as it is seldom recorded on the building. Approaches may need to be made to the following sources, who may also have valuable information, possibly (but rarely!) including some or all of the structural information being sought (Fig. 3.2):

- present and past building owners and tenants, and their professional advisers (architects, surveyors, building managers, agents, solicitors, etc.);
- the original and subsequent building contractors and sub-contractors, concrete element suppliers, etc.);
- the building control authority;

- contemporary technical journals;
- specialist libraries with extensive archive collections, notably the Institution of Civil Engineers, the Institution of Structural Engineers and the Royal Institute of British Architects, the ICE library holds the Concrete Society Archive (see Appendix A);
- other archive collections, including the Public Record Office (public structures, including much on Government buildings, railways, etc.), Railtrack (railway structures still in use), British Waterways, etc.

The designer of the structure was not, and still is not always, an independent consulting engineer employed by the building owner. All reinforced concrete structures have been (or should have been) 'engineered', i.e. designed by a competent person making calculations and producing drawings, specifications and so on from which the structure was built. Much of this work is originated by consulting engineers, but they have never had a monopoly.

In the early years of reinforced concrete, as explained in Chapter 2, the overall design was usually carried out by the specialist concrete firms in-house. This, although on a reduced scale, has continued to the present day. For example, after World War II many shell and barrel-vault roofs were designed and built by companies such as Twisteel Reinforcement Ltd (later GKN Reinforcements Ltd) and BRC (British Reinforced Concrete Engineering Co Ltd) (Anchor, 1996). Many of the precast large-panel housing systems adopted in the 1960s were designed by the companies offering them, although often they in turn employed consulting engineers. Similarly, multi-storey car parks were often offered as a 'design and build' package by companies such as Bison Ltd.

In addition, even 'one-off' structures frequently incorporate standard structural elements, usually precast, such as flooring beams and joists, lintels and stair flights. The building designer (architect, engineer or surveyor) would prescribe loadings and typically size the elements using the concrete company's brochure. The company would then produce the necessary drawings and calculations, these often being standardised and requiring only project-specific information to be added by hand before submission for approval by the designer and the building control authority.

A further complication may arise once the original designer's name has (with luck) been traced. Many consultants and contractors have gone out of business, or have merged with or been acquired by others, once or more often. Their name may well be untraceable. In such cases, the Association of Consulting Engineers, Companies House, building trade associations or the libraries mentioned above might be able to help (see Appendix A for contact details of such organisations).

Against this, it is a consolation that many long-established firms are still active, and have carefully maintained archives. Among such consultants who have been established for at least half a century are Mouchel (designers of the many Hennebique structures) Hurst Peirce and Malcolm, Arup, Oscar Faber and F.J. Samuely (see Chapter 2).

Approaches for structural information should accept that it is commercially valuable. A charge may be made for access to it, and conditions may be imposed on its use.

Generic information

Generic information includes:

- contemporary building regulations, codes of practice and standards;
- contemporary textbooks and technical journals;
- contemporary trade literature;
- historical engineering studies, including Sutherland *et al.*, 2001, which has extensive bibliographies for further guidance and help, and a Construction Industry Research and Information Association review of post-tensioning systems (Andrew and Turner, 1985);
- retrospective guidance on appraisal and/or treatment of particular materials, building types, construction techniques, and the like, including many Building Research Establishment publications, such as Bate (1984) and the BRE (1994) for high alumina cement structures, Currie *et al.* (1987) for large-panel precast concrete structures, and the BRE (1984) for concrete houses.

These are generally available in the specialist libraries noted above. Other fruitful archives are to be found in the British Library's Science Reference and Information Service.

Detailed investigation

The investigation will typically include some or all of the following:

- identifying and locating structural elements to establish the overall structural form;
- establishing the type and disposition of reinforcement in elements and connections;
- establishing the presence, thickness and weight of non-structural elements, including partitions, cladding and glazing, finishes and services;
- locating defects and identifying their cause(s) (see Chapter 4).

However good the documentary information obtained (and it is rarely comprehensive!), it is essential to investigate the structure as it stands. Seldom do drawings, even those noted as 'as-built' records, show exactly what was built, and even if they do the building will probably have been altered during its lifetime. Furthermore, drawings do not indicate the present condition.

All the same, the scale and extent of the investigation will be influenced by the amount and quality of documentary information to hand. Clearly the availability of drawings will be of great value: it is both tedious and disruptive to have to hack away screed and plaster to establish the presence and thickness of concrete elements, let alone to then attempt to map the reinforcement.

Locating and identifying reinforcement

If reinforcement drawings are not available, then for a full appraisal it will be necessary to consider investigative methods. If the building is to be refurbished there is little difficulty in the approach of 'knocking it about' with a hammer and cold chisel to expose the reinforcement, and using a cheap metal detector to locate the presence of steel. The bar cover, diameter and spacing can then be measured. However, this is slow and disruptive, and might be unacceptable, especially on exposed 'architectural' concrete surfaces.

The quicker and less disruptive alternative is the use of non-destructive testing (NDT) techniques, and the various methods are outlined in Chapter 4 in the section on condition survey techniques. The cheap metal detector, as sold for locating domestic pipes or nails, is of no use here. There are various makes of 'covermeters' available for purchase or hire, which generally make use of the response of an electromagnetic field to the presence of steel to register a reading. As their name implies, these are intended to detect an individual bar and then measure its concrete cover; as such, they are widely used to check compliance with the specification during construction. Most models also have provision for estimating the bar diameter, although this usually calls for an experienced operator, and results should be spot-checked frequently by opening-up.

At least one model now available uses microprocessing to 'scan' areas of the concrete in two orthogonal directions, and then delivers an image of the reinforcement pattern, showing spacing and cover. It can also indicate probable bar diameter. The images can be downloaded to a computer.

One further NDT method is the use of radar. This is a specialist and fairly expensive technique. Experienced personnel are needed to interpret the radar plots. Once again, the results should be spot-checked frequently.

All of these machines have a relatively shallow range, so that they locate only the bars nearest the concrete face being examined. In general, too, they are unable to identify the type of steel, i.e. whether it is mild or high-yield, which is an important issue for strength assessment. They may also have difficulty in dealing with stainless steel, which typically is non-magnetic. Once again, opening-up will be needed to support results from the NDT approach.

Present-day reinforcement generally takes one of three forms: plain round bars of mild steel; ribbed or deformed high-yield bars; or 'fabric' or orthogonal sheets of high-yield strength mesh comprising round plain or ribbed bars spot-welded together at intersections. These can be distinguished by eye, and related to current British Standards for strength and other properties.

Many other bar profiles have been used for reinforcement, particularly in the early years of the twentieth century when the various proprietary systems were in use (see Chapter 2). Most of these were of mild steel, although some were cold-drawn, which resulted in higher strength (at the expense of reduced ductility). Identification of the reinforcement, and hence the system, will be aided by Figs. 2.8–2.10, which show some early bar types. Further assistance in identification will be found in the archives of the libraries of the Institution of Civil and Structural Engineers mentioned above, together with more detailed descriptions of the systems and their characteristics.

Investigating foundations and groundworks

Foundations and other groundworks, such as basements, present a particular if obvious problem—how to get at them! Trial pitting is a common and traditional solution for shallow strip and pad foundations, but the heavier work associated with larger structures and/or poorer ground conditions, such as piles, rafts and diaphragm walling, will usually prove impractical to expose at reasonable cost. This emphasises the importance and potential value of a thorough documentary search.

BD21/01 (Highways Agency *et al.*, 2001) offers some comfort in cases where information on such construction is limited, particularly where loads will not be increased and load paths are not being altered. Paragraph 8.5 states:

> If a foundation, retaining wall or a substructure shows no signs of distress, if there is no evidence of scour either externally or internally, and if no significant increases in load are envisaged, then the foundation, retaining wall

or sub-structure [*sic*] may be assumed to be adequate and no further assessment is necessary.

Although this guidance is intended for bridges (hence the reference to scour, which is applicable only to structures founded in waterways), it is arguably equally applicable to buildings, and is a welcome piece of practical engineering common sense enshrined in an official document.

Testing: general

Testing is commonly undertaken as part of the detailed investigation stage. Commissioning testing may be thought to be 'making a good start', evidence of 'something being done', but before rushing into this it is prudent to think through the questions considered below.

What information is the testing required to provide?

Here, strength of concrete and reinforcement, and perhaps the ductility of the reinforcement, are the properties to be considered for testing prior to structural assessment. Testing as part of a 'health check' on deleterious materials or degradation is discussed in Chapter 4 in the section on sampling. Structures built since the 1960s may also warrant sampling to establish the type of cement or cement replacement material(s) used in the concrete.

Is testing really necessary?

For a first assessment, particularly if it is required quickly, it may be adequate to assume values for strength given later in this chapter. Alternatively, or as a complementary check, concrete strength may be estimated using a rebound hammer or ultrasonic pulse velocity measurements (Bungey, 1992). Both methods can give a reasonable indication of strength in skilled hands.

Only if the assessment shows that the structure falls significantly short on strength would testing of material samples then be warranted. Even then, testing would not be justified if the assessment showed, for example, that a scheme substantially to increase loads on an early twentieth century concrete structure demanded a concrete strength of $60\,\text{N/mm}^2$ or a reinforcement strength of $600\,\text{N/mm}^2$. Such high strengths will not be found! Here, one would proceed straight to thoughts of strengthening (or try to persuade the client to abandon the idea).

What will be the outcome of 'good' or 'bad' test results?

It is important to consider what consequences will follow from the results obtained before carrying out the testing. This is particularly relevant for load testing, discussed below. A 'good', that is better than required, test result for strength will usually provide comfort and assurance that the structure can perform satisfactorily, and is worth paying for. A 'bad' result, i.e. lower than required, will typically lead to a decision to abandon a scheme, such as for increased loading as outlined above, or to undertake strengthening work.

Having given these matters thoughts, some testing may be pre-planned as an integral part of the work. Other testing may be called for in response to findings as the investigation proceeds. There are numerous testing houses which can provide the required assistance, including on-site attendance, cutting and removal of samples, testing and reporting.

It should not be necessary to add that the removal of materials for testing must not weaken the structure. Accordingly, samples must be taken from lightly stressed locations or elements. The tale is told, salutary even if only apocryphal, of the newly built columns whose concrete was of suspect quality following a low cube strength result for the batch from which they were cast. Consultant, contractor and ready-mixed concrete supplier each commissioned core samples from their respective testing houses to obtain a better idea of actual strength. The resulting strength figures were judged acceptable. Unfortunately, all the cores had been taken from a single column which now looked like a piece of Gruyère cheese and was fatally weakened. It had to be demolished and rebuilt!

Sampling for testing should generally be on a 'random' basis of location, subject to two considerations. Firstly, it is essential that where test results are to be considered together for determining characteristic strengths (see below), then the samples must be taken from the same population. Thus, it would be inappropriate to base concrete strength on results combined from precast cladding panels, *in situ* walls and proprietary precast floor joists, since these have come from three different sources and were probably specified with different strengths initially.

Care is also needed in regard to *in situ* construction, where the element strengths may also have been specified with different values. It was common practice, certainly in the 1960s–1980s, to specify columns of a higher-grade concrete in commercial developments and high-rise blocks, in order to reduce the 'footprint' of the columns and hence maximise available floor area. In rare (and fortunately usually well-documented) cases a concrete column was reinforced, indeed virtually replaced, by the use of a solid steel billet.

The second consideration for sampling location is where only a small part of the structure is in doubt. If, for instance, only one bay of a floor is to carry enhanced loading, and only the four columns supporting it are questionable, it is logical to focus on these columns for sampling purposes (not forgetting the caveat above on weakening the structure).

Testing concrete for strength

Concrete to be tested for strength is normally removed from an existing structure by 'coring' using a diamond-tipped hollow coring drill. Cores as small as 38 mm diameter can then be tested by crushing, although wherever possible a larger diameter core (50–150 mm) should be taken as this will avoid the effect of a large piece of coarse aggregate. Such cores should be located clear of any nearby reinforcement in order to avoid damage to the structure and also to avoid having steel in the core, which again will distort the strength results for the concrete. Core-holes are usually filled with a fine concrete to reinstate the original strength of the element.

The core strengths obtained by crushing can be factored to give an equivalent 'cube' strength, this being the standard benchmark of concrete strength for design and assessment (BS 6089, 1981; Concrete Society, 1987).

In structures built before the mid-1980s, the equivalent cube strength as determined from testing may be considerably higher than that originally specified. The reason for this is that cement was not as finely ground as it is today, and consequently the concrete did not gain strength as rapidly as it does in modern mixes, which achieve full strength earlier. This allows faster stripping of formwork and hence faster progress on-site. The mid-twentieth century codes of practice CP 114 (1957) and its successor CP 110 (1972) allowed the cube strength 1 year after construction to be taken as up to 25% greater than it was at the time of construction.

A further reason for higher-than-expected cube strengths in more recent structures is that where ready-mixed concrete, in particular, was used, the batching plant would often deliberately provide a mix designed for a higher strength. This was intended to ensure that the concrete would not give rise to arguments when the routine cube test results were issued and subjected to statistical checks against the specification requirements.

The cube strengths obtained from testing may be used in preference to the original specified values once they have been processed to give a 'characteristic strength' as defined in BS 8110. This is given by the following formula in DD ENV 1991-1 (1996):

Table 3.1   Relationship between k and the number of test results.

Number of samples	3	4	5	6	8	10	20	∞
Value of k		3.37	2.63	2.33	2.18	2.00	1.92	1.76	1.64

$$f_{char} = f_{mean} - k \cdot \sigma$$

where f_{char} is the characteristic strength, f_{mean} is the mean strength obtained in tests, k is a numerical constant given in Table 3.1 and σ is the calculated standard deviation. It can be seen that typically six to ten tests will give a fair estimate of the characteristic strength. More will not greatly improve the figure; fewer will push it downwards.

In view of the notional possibility of a negative value of f_{char} being obtained using this formula, a log–normal form is preferable to ensure that the value will be positive, as follows:

$$\log f_{char} = \text{mean}(\log f) - k(\text{standard deviation of } \log f)$$

in which f stands for the individual test results.

BD21/01 (Highways Agency *et al.*, 2001) suggests that for bridges the characteristic strength of pre-1939 concrete may be taken as $15\,N/mm^2$. This may sound low when strengths today are routinely specified as $30\text{--}40\,N/mm^2$, but the weakest 1:2:4 Ordinary Grade mix strength specified in the 1933 DSIR Code was indeed only $2200\,lb/in^2$ or $15.5\,N/mm^2$. So an assumed strength of $15\,N/mm^2$ may not be unrealistic for a first assessment of pre-1939 structural concrete generally, although subsequently testing for strength would be worthwhile. Of course, an effort should be made to locate the records of construction, which would give documentary proof of the specified and achieved concrete strengths. Where a cube strength is seen to have been specified, this can be taken as the characteristic strength.

Testing reinforcement for strength

Reinforcement strength is usually obtained by a standard tensile test. BD21/01 (Highways Agency *et al.*, 2001) suggests that for bridges the characteristic (yield) strength of pre-1961 reinforcement be taken as $230\,N/mm^2$, a figure which is appropriate for mild steel (Highways Agency *et al.*, 1997). For structures of a later date, the appropriate values from the contemporary code of practice may be used. Typical figures are: for mild steel $250\,N/mm^2$; for high-yield steel not less than $410\,N/mm^2$

(1961–1980), $425\,N/mm^2$ (1980–1983) or $460\,N/mm^2$ (after 1983); for mesh $410\,N/mm^2$ (1961–1972), $485\,N/mm^2$ (1972–1983) or $460\,N/mm^2$ (after 1983) (Bussell, 1996).

Testing prestressing steel for strength

Here, the advice of Mr Punch is best—'don't'! Prestressing steel is tensioned to a high stress level, and even when bonded to the surrounding concrete it can 'whip' and cause serious injury if cut. In addition, prestressed structures generally do not contain lightly stressed sections of steel, and require longer bond and anchorage lengths in view of the higher steel stress, so that the removal of steel may well compromise strength. The strength of proprietary prestressing steel can be obtained from contemporary codes of practice and trade literature (Institution of Structural Engineers, 1951; Andrew and Turner, 1985; CP 115, 1959; CP 110, 1972).

Recording

Records must be made as the investigation proceeds. These should include sketch drawings, notes and photographs (still and video). Modern aids to effective working include the cassette recorder for making notes, and digital cameras to allow the direct transfer of images to the computer for study and inclusion when preparing an appraisal report. However, making useful sketches and notes is an active mind-engaging task, involving thought, if everything of relevance is to be noticed and recorded. In this respect it can be argued that the 'press and talk' and 'point and shoot' approach, using recorder and camera, is more likely to miss important observations.

Assessment of structural condition

Assessment involves looking at the structure as it is, taking account of its actual construction and its present (and future) condition. In this respect it differs fundamentally from the design of a new structure.

Assessment compared with design

Design involves conceiving a structure that can demonstrably carry loads to the ground. It must be safe, strong, stable and stiff. The routes by which the loads reach the ground (the 'load paths') are usually simple and direct.

For example, a slab is carried by a beam, supported by a column whose load is delivered to the ground through its foundation. The concrete designer makes what is effectively a mathematical model of the structure. Structural analysis calculates the forces and movements produced in this model by the loads to be carried, and uses an accepted design method to determine element sizes and reinforcement requirements based on appropriate material strengths. The 'factor of safety' used in the design takes account of possible modest variations in element sizes, material strengths, position of reinforcement and so on that will occur during construction. The design is translated into drawings, specifications, bending schedules and the like that describe what is to be built.

Assessment, in contrast, involves looking at the reality of the built structure as it today. Unless it has totally collapsed, it has found a way of standing up. This may well bear little or no relation to the assumptions about behaviour made by the designer, and the stresses at any point are almost certainly *not* what the designer calculated. This apparently baffling state of affairs can be made clearer by an example.

A short-span deep beam, supported at either end, might have been designed to carry its own weight on simple 'beam theory', which assumes that it bends like a timber floor joist. The bending effect, or bending moment, varies parabolically along the span, being zero at the supports and rising to a maximum at mid-span. In a beam or joist, the strain on the section due to bending varies from a maximum in tension at the bottom, which is being stretched apart, reducing to zero at or near the mid-height of the section, and then changing to compression above this. The greatest compressive strain is at the top edge of the section, which is being squeezed together along its axis. The reinforcement was calculated and provided to resist the maximum bending tension in the bottom of the mid-span, assuming this mode of behaviour. According to this model, the bending decreases towards the supports, so that the stress in the tensile reinforcement will also reduce. On the basis of this theory, the tensile stress in the reinforcement dwindles to zero at the supports.

An equally valid way of visualising the working of this beam is as a bowstring arch, with a parabolic arch of compressed concrete acting within the beam section. This 'virtual' arch is at its greatest height at mid-span, and drops downwards towards the supports. According to this model, the profile of this arch matches that of the bending moment, so that the tension in the bottom reinforcement is now constant—indeed it is acting as a tie to hold the ends of the arch and prevent them from spreading. Because of this, it is important to ensure that the bars are well-anchored into the concrete at both supports.

Both models are valid, despite offering two quite different approaches to representing the structural behaviour of the beam. Both provide a 'safe' design method provided that the implications of the particular behaviour mode are acknowledged in the design and detailing of the structure.

Use of codes of practice in assessment

Structures designed to earlier, usually more conservative, regulations or codes of practice may nevertheless be assessed using present-day codes. The materials in the structure are, after all, unaware of how they are supposed to be working! This approach has advantages when assessing a structure, particularly for alteration purposes: it will often allow a somewhat higher capacity to be adopted. This is because design philosophy has evolved as testing and thinking have improved our knowledge of the structural behaviour of reinforced concrete in service and up to failure.

UK regulations and codes of practice prior to the 1950s (see Chapter 2) all based their recommendations for design on the so-called 'elastic permissible stress' method. This considered the behaviour of the structure under 'working' (i.e. actual) loads. It adopted allowable stresses on the concrete and the reinforcement related to the crushing strength of the concrete and the breaking strength (later the yield strength) of the reinforcement. It also linked the stresses in the two materials by assuming 'strain compatibility', i.e. that both materials, being bonded together, stretched or shortened by the same amount when loaded. As steel is about 15 times stiffer than concrete, this limited the steel stress to about 15 times that in the concrete. This often controlled the design, so that using higher-strength reinforcement gave no benefits in reduced steel requirements, which often afforded simpler construction and hence reduced costs.

Tests in the 1930s by Oscar Faber and others on reinforced concrete columns showed that this model for behaviour was unrealistic. Drying shrinkage and subsequent concrete 'creep' (relaxation under sustained loading) threw considerably more load onto the main column-reinforcing bars. Further research showed that the behaviour of beams and slabs at failure loading was significantly different from that in the 'elastic' model. Similarly, columns and walls failed only after mobilising the combined strength of the concrete and the reinforcement. This was reflected by changes to the UK reinforced concrete code from the 1950s, although design at this time was still for the 'working' load condition (Bussell, 1996).

This changed with the issue of BS CP 110 in 1972, a so-called 'limit state' design code for structural concrete. In this, design for strength considered the 'ultimate limit state of failure', for which loads were factored up by 'partial load factors' (e.g. 1.4 for dead loading such as self-weight and finishes, and 1.6 for imposed loading, i.e. user loads). Similarly, the material strengths assumed at failure were to be reduced by 'partial material factors' (1.5 for concrete in compression, reflecting its more variable nature because it is made in batches, and 1.15 for reinforcement, i.e. a more 'controlled' product).

In fact, the partial factors were both made up of several independent variables based on statistical theory, although the final figures were calibrated so that designs to this code would have a similar level of safety (and cost) to those based on its predecessor (Institution of Structural Engineers, 1996). (BS CP 110 also dealt with design for the 'serviceability limit state', which was principally to limit deflections under load, and to ensure that cracking under load would not jeopardise durability by aiding corrosion of the reinforcement.)

Approach to assessment

A logical and systematic approach is essential in all stages of the appraisal. A typical sequence for an assessment is shown in the flow-chart in Fig. 3.3 (BRE, 1991). Generally, it will be necessary to carry out a loading 'run-down', beginning with gravity loading, including dead and imposed loads and snow, and then accumulating these onto slabs and beams, down columns and walls, and finally to foundations and the ground. Imposed load reductions permitted by current loading codes may be applied. Wind loading must also be assessed.

Here, and particularly in the subsequent analysis of the structure to determine the forces acting, the assessment will often be iterative, beginning with a simple model to obtain a 'feel' for its adequacy before refining it, if necessary, to reach a firmer conclusion. There is much to be said for at least the first calculations being done by hand rather than by computer, and concentrating thought on the actual structure rather than the mathematical model used in the design.

An analysis of the structure will be made for strength. In the current 'limit state' approach, as described above, the loads and material strengths will be applied with partial factors being used. There is a precedent from assessments carried out on high alumina cement elements for using a reduced dead load factor of 1.2 where the actual thickness and weight of dead load elements has been determined (Department of the Environment and the Welsh Office, 1975). This acknowledges the reduced uncertainty on loading as a result of such a check.

Stability and the effects of lateral loading must also be considered: can the structure resist the applied loading without falling over, and what are the forces due to wind and other lateral loading?

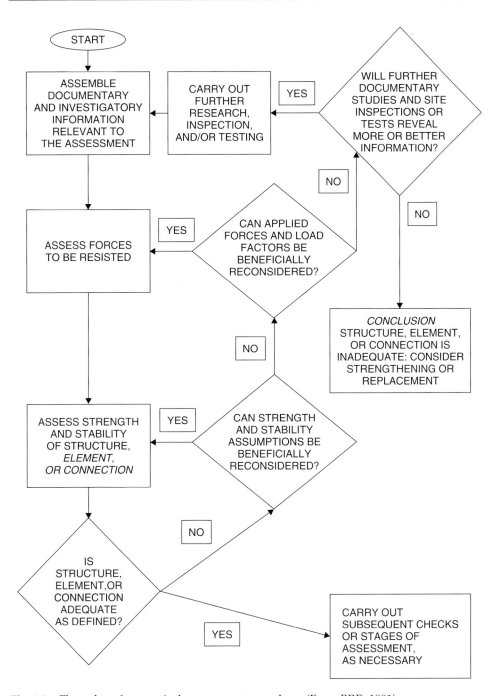

Fig. 3.3 Flow-chart for a typical assessment procedure. (From BRE, 1991).

Finally, for structures of a certain height (currently five or more storeys), Building Regulations Requirement A3 applies (Department of the Environment *et al.*, 1994). This concerns the avoidance of progressive or 'disproportionate' collapse, and was originally introduced after the Ronan Point collapse in 1968 (described in Chapter 2). It is intended to ensure that the structure is generally robust, and in particular that in the event of, say, a gas explosion or vehicle impact, the structure will not suffer collapse disproportionate to the cause. In practice, it is generally recognised that certain forms of construction, notably steel frames and *in situ* concrete frames, are inherently robust. However, large-panel precast concrete panel construction and 'hybrid' structures using a variety of materials and forms (for example brick or block walls used with timber or precast concrete floors) require more explicit attention to joints and connections to ensure their integrity. Such a check may be needed, particularly when a material change of use is proposed (BRE, 1991).

Conclusions from the assessment

The calculated forces acting can now be compared with the element capacities. This will produce one of three results:

- the structure has adequate strength;
- the structure, or part of it, is grossly inadequate;
- the structure, or part of it, is 'nearly' adequate.

The first result (obviously the best) requires only careful checking to ensure that nothing has been overlooked before proceeding to the report (see below). Many structural engineers and building control authorities would regard a few per cent 'overstress' as adequate for superstructure elements. There is a wide precedent for accepting up to 10% increased loading on existing foundations if there is no evidence of ground movement or related distress such as settlement.

The second result may cast doubt on the analysis as much as on the structure, especially if the structure is standing with no evident signs of distress. In such cases the analysis assumptions, and indeed loadings and calculations generally, should be carefully reviewed for gross errors and also to see if a more rigorous analysis is appropriate. If strengths have been assumed, testing to determine actual strengths can only help the appraisal, and should be undertaken. Every effort should be made to ensure that the assessment has been as close to 'reality' as possible. As a last resort, load testing (see below) may be considered.

The third result, 'nearly' adequate, can be more problematical than the other results. 'Nearly' is a debatable definition, but here it is taken to be in the range of, say, up to 20% understrength. Again, if strengths have been assumed, testing is probably now warranted. Certainly the assessment should be reviewed to see if assumptions and analyses could be refined to allow the structure to be classed as adequate.

Particular effort to reach this conclusion is warranted in the case of listed structures, especially those having exposed architectural concrete finishes which could be visually compromised by the intervention necessary for repair, strengthening or replacement.

Serviceability and durability

The serviceability and durability of the structure are also considerations alongside strength and stability. The structure should be stiff enough to work without deflections and movements that impair its use. Reinforced concrete, properly designed and built, is a stiff material, but particular circumstances may need careful investigation. Examples are the installation of rolling book stacks, which on a lively floor may start moving on their own, and delicate scientific equipment whose operation may be flawed by vibrations from foot traffic or other causes. Repair and maintenance to reinstate and ensure durability in the future is covered in Chapters 5 and 6.

Consideration of work needed

If the structure has finally been judged to be inadequate, then some action must be considered.

Load testing

Load-testing is one possibility, but it raises the issue of the selection and extent of the areas to be tested. It is a slow and expensive process, with no guarantee that the structure will 'pass'. As with testing in general (considered above), the consequences of such a 'fail' result should be considered before commissioning load testing. It may, in any event, be cheaper to carry out remedial work than to carry out a load test!

Load-testing should be fully specified (Institution of Structural Engineers, 1989). Typically, a floor or roof area is loaded up to 1.25 times the total service load it is to carry. The 'pass criteria' are usually that the structure can carry this load without exceeding a pre-defined limit of deflection (often derived from the relevant code of practice limit), and retaining no more than a considerably smaller residual deflection after unloading. The latter check is to ensure that the reinforcement has not yielded at this load level. Temporary works during the testing should include fail-safe back-propping as a safety precaution should the structure fail locally, and the means of applying the test load and of measuring the deflections. These measures should also respond to safety issues by not requiring test personnel to stand on or under the area being tested.

Repair, strengthening and replacement

Repair is intended to restore the strength of the structure to a pre-existing state, whereas strengthening enhances strength and replacement substitutes a newer, stronger or sounder element.

In all cases, thought must be given to the temporary works required. This includes ensuring that any temporary reduction in strength during the work is considered, and suitable measures are provided. Repair generally is considered in Chapter 5.

Strengthening can take one of several forms, including:

- composite plate bonding to enhance tensile bending resistance (see Chapter 5);
- chasing-in of additional reinforcement;
- casting of a bonded screed on top of a slab or beam to form a composite section;
- insertion of additional supporting structure (beams, columns, etc.) to shorten the span of slabs and beams, taking care to consider whether this might cause cracking where the supported elements have inadequate or no reinforcement in the face newly put into tension (e.g. propping the tip of a drooping cantilever slab which has no bottom reinforcement);
- encasing of a column or 'facing' of a wall with additional concrete and reinforcement.

Replacement involves the removal of the inadequate element and construction of a new element of adequate strength. This may be the only practical solution following bomb damage or vehicle impact. Invariably it is a major task—particularly if the element to be replaced is a column or wall, necessitating shoring-up to relieve the element of loads from above. All such work calls for experience and specialist skills in planning, design and implementation (see Chapter 5; also: Currie and Robery, 1994; Concrete Society, 1984).

Reporting and recommendations

No appraisal is complete until it has been recorded. Clearly, the client will want to have a written statement describing what has been done and what has been concluded. For a modest exercise, the report might be in the form of a letter. On larger studies, the final report might run to several volumes, although interim reports might again be in letter form. Such an interim report might be drafted at the stage when strength testing is being recommended, or when the possibility of a load test is being discussed.

Whatever its form, a report should always seek to present the facts, and to keep these clearly separate from the opinions, conclusions and recommendations drawn from this factual evidence. Experienced report-writers will need little advice on this aspect, while the less experienced can gain much from general guides on the writing of technical reports (for example Scott, 1984). However, the following order of contents is offered as a typical and logical basis for a report that (suitably expanded or contracted) is applicable to any scale of reporting:

- executive summary or synopsis;
- brief for the appraisal;
- documentary investigations (with lists of sources located, names of parties involved, and documents reviewed, preferably contained in one or more appendices; copies of drawings and other documents obtained might warrant inclusion in additional volumes);
- structural description (its construction and related factual information);
- investigations and observations (detailed inspection records, e.g. text, drawings, photographs, etc., usually to go in appendices, although a selection of these within the main report can add flavour to the narrative and also assist the reader);
- sampling and testing (records to go in appendices; reports by other parties such as a testing house, etc., to be attached or identified);
- description of assessment procedure;
- discussion of assessment findings;
- conclusions—the adequacy or otherwise of the structure (often combined with the following section of recommendations);
- recommendations (further work or testing; remedial measures, e.g. repair, strengthening, replacement; other suggestions for achieving adequacy such as modifying proposed existing or future use);
- appendices (contents as suggested above, as appropriate to the particular case).

References

Anchor, R.D. (1996) Concrete shell roofs, 1945–1965. *Proceedings of the Institution of Civil Engineers: Structures and Buildings* **116**:381–389.

Andrew, A.E. and Turner, F.H. (1985) *Post-Tensioning Systems for Concrete in the UK: 1940–85, Report 106.* Construction Industry Research and Information Association, London.

Bate, S.C.C. (1984) *High Alumina Cement Concrete in Existing Building Structures. Report BR235.* HMSO, London.

BRE (1984) *The Structural Condition of Prefabricated Reinforced Concrete Houses Designed Before 1960. Information Paper 10/84.* Building Research Establishment, Garston.

BRE (1991) *Structural Appraisal of Existing Buildings for Change of Use. BRE Digest 366.* Building Research Establishment, Garston.

BRE (1994) *Assessment of Existing High Alumina Cement Concrete Construction in the UK. BRE Digest 392.* Building Research Establishment, Garston.

BS 6089 (1981) *Guide to Assessment of Concrete Strength in Existing Structures.* British Standards Institution, London.

Bungey, J.H. (1992) *Testing Concrete in Structures: A Guide to Equipment for Testing Concrete in Structures. Technical Note 143.* Construction Industry Research and Information Association, London.

Bussell, M.N. (1996) The development of reinforced concrete: design theory and practice. *Proceedings of the Institution of Civil Engineers: Structures and Buildings* **116**:317–334.

Concrete Society (1984) *Repair of Concrete Damaged by Reinforcement Corrosion. Technical Report No 26.* Concrete Society, Slough.

Concrete Society (1987) *Concrete Core Testing for Strength. Technical Report No. 11.* Concrete Society, Slough.

Concrete Society (1990) *Assessment and Repair of Fire-Damaged Concrete Structures. Technical Report No. 33.* Concrete Society, Slough.

CP 114 (1957) *The Structural Use of Reinforced Concrete in Buildings.* British Standards Institution, London.

CP 115 (1959) *The Structural Use of Prestressed Concrete in Buildings.* British Standards Institution, London.

CP 110 (1972) *The Structural Use of Concrete.* British Standards Institution, London.

Currie, R.J., Reeves, B.R., Moore, J.F.A. and Armer, G.S.T. (1987) *The Structural Adequacy and Durability of Large Panel System Dwellings. Report BR107. Part 1. Investigations of construction. Part 2. Guidance on appraisal.* Building Research Establishment, Garston.

Currie, R.J. and Robery, P.C. (1994) *Repair and Maintenance of Reinforced Concrete. Report BR254.* Building Research Establishment, Garston.

DD ENV 1991-1: Eurocode 1 (1996) *Basis of Design and Actions on Structures. Part 1. Basis of Design (Together with National Application Document).* British Standards Institution, London.

Department of the Environment and the Welsh Office (1975) *Building Regulations Advisory Committee: Report by Sub-Committee P (High Alumina Cement Concrete).* Department of the Environment and the Welsh Office, London.

Department of the Environment (now Department of the Environment, Transport and the Regions) and the Welsh Office (1994) *The Building Regulations 1991: Approved Document A—structure*. HMSO, London.

Health & Safety Commission (2001) Managing Health and Safety in Construction: Construction (Design and Management) Regulations 1994. *Approved Code of Practice*. HSE Books, Sudbury.

Highways Agency, Scottish Executive Development Department, The National Assembly for Wales, and Department for Regional Development (Northern Ireland) (2001) *Design Manual for Roads and Bridges. Vol. 3. Highway Structures. Section 4. Assessment. Part 3. The Assessment of Highway Bridges and Structures, BD21/01*. The Stationery Office, London.

Institution of Structural Engineers (1951) *First Report on Prestressed Concrete*. Institution of Structural Engineers, London.

Institution of Structural Engineers (1989) *Load Testing of Structures and Structural Components*. Institution of Structural Engineers, London.

Institution of Structural Engineers (1996) *Appraisal of Existing Structures*, 2nd edn. Institution of Structural Engineers, London.

Jones, B.E. (ed) (1920) *Cassell's Reinforced Concrete*, 2nd edn. Waverley, London.

Scott, B. (1984) *Communication for Professional Engineers*. Thomas Telford, London.

Sutherland, J., Humm, D. and Chrimes, M. (eds) (2001) *Historic Concrete: Background to Appraisal*. Thomas Telford, London.

Watt, D. (1999) *Building Pathology: Principles and Practice*. Blackwell Science, Oxford.

Further reading

BRE (1993) *Concrete Cracking and Corrosion of Reinforcement. BRE Digest 389*. Building Research Establishment, Garston.

BRE (1995) *Static Load Testing: Concrete Floor and Roof Structures Within Buildings. Digest 402*. Building Research Establishment, Garston.

BRE (2000) *Corrosion of Steel in Concrete. BRE Digest 444. Part 1. Durability of Reinforced Concrete Structures. Part 2. Investigation and Assessment. Part 3. Protection and Remediation*. Building Research Establishment, Garston.

Concrete Society (1992) *Non-Structural Cracks in Concrete. Technical Report No. 22*, 3rd edn. Concrete Society, Slough.

Health & Safety Executive (1990) *Evaluation and Inspection of Buildings and Structures. Guidance Note G58*. Her Majesty's Stationery Office, London.

Institution of Structural Engineers (1991) *Guide to Surveys and Inspections of Buildings and Similar Structures*. Institution of Structural Engineers, London.

Pullar-Strecker, P. (1987) *Corrosion Damaged Concrete: Assessment and Repair*. CIRIA/Butterworth, London.

Chapter 4

The Identification and Assessment of Defects, Damage and Decay

John Broomfield

Introduction

There are a number of causes of deterioration in concrete buildings and structures. Even when they are adequately built, properly used and well maintained the environment will affect a structure and components will wear out. The largest single cause of deterioration in reinforced-concrete structures is corrosion of the reinforcing steel. In addition, there are a number of deterioration processes that attack the concrete directly, some from within, such as alkali–silica reactivity, and some from without, such as freeze–thaw damage. Others are related to initial construction problems or subsequent use or abuse of the structure.

This chapter summarises the major causes of defects, damage and decay in concrete buildings and structures. Any attempt to remedy problems must start with a thorough understanding of the cause and extent of the deterioration. As discussed in Chapter 3, it is therefore essential that a detailed investigation is carried out as part of the appraisal process, the results are interpreted, and the repair options fully evaluated to ensure that the right repair option is selected for the building and its owner. This chapter is therefore an essential precursor to Chapter 5, which deals with repair options. The main causes of concrete deterioration are summarised in Fig. 4.1, and these are discussed in detail in the following sections.

Corrosion of steel in concrete by carbonation and chlorides

One of the many fascinating properties of concrete is its relationship with the steel embedded in it. Historically, attempts were made to reinforce concrete with metals such as bronze. The resulting failures were due to the difference in thermal expansion coefficients between the bronze and the concrete. Steel has a similar enough thermal expansion coefficient to concrete to avoid damage. However, steel corrodes in a moist oxygenated atmosphere. Despite its porosity and the natural moisture level of

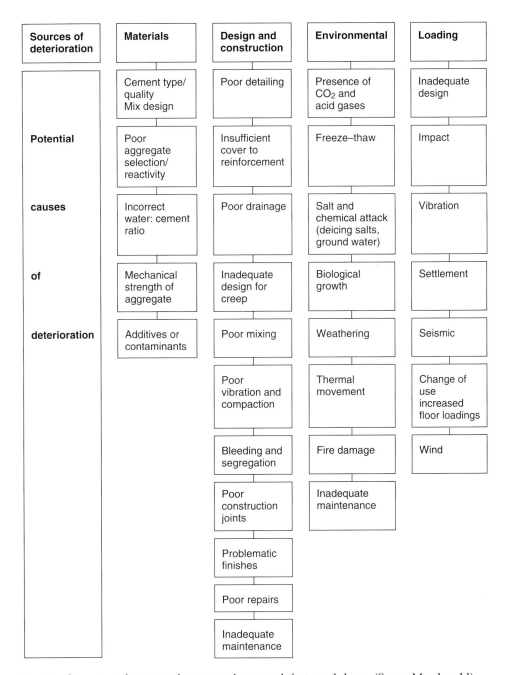

Sources of deterioration	Materials	Design and construction	Environmental	Loading
	Cement type/ quality Mix design	Poor detailing	Presence of CO_2 and acid gases	Inadequate design
Potential	Poor aggregate selection/ reactivity	Insufficient cover to reinforcement	Freeze–thaw	Impact
causes	Incorrect water: cement ratio	Poor drainage	Salt and chemical attack (deicing salts, ground water)	Vibration
of	Mechanical strength of aggregate	Inadequate design for creep	Biological growth	Settlement
deterioration	Additives or contaminants	Poor mixing	Weathering	Seismic
		Poor vibration and compaction	Thermal movement	Change of use increased floor loadings
		Bleeding and segregation	Fire damage	Wind
		Poor construction joints	Inadequate maintenance	
		Problematic finishes		
		Poor repairs		
		Inadequate maintenance		

Fig. 4.1 Sources and causes of concrete damage, defects and decay (Susan Macdonald).

concrete exposed to the outdoor atmosphere, steel does not necessarily corrode once it is encased in concrete for reasons discussed in the following section.

There are two major mechanisms for the corrosion of steel in concrete that do not require the degradation of the concrete before the steel is attacked. The first of these is carbonation and the second is chloride attack.

Carbonation

Reinforcement corrosion is prevented by the alkali content of the concrete. During the hydration process calcium, sodium and potassium hydroxides are formed, which dissolve in the pore water of the concrete to form a very alkaline solution of around pH 12–13.5. At this level, steel forms a very thin, protective oxide known as a passive layer. This is self-sustaining and maintaining; it is far better than synthetic or metallic coatings that deteriorate or are consumed. The passive layer will sustain and maintain itself indefinitely as long as the alkalinity stays above about pH11 without contamination.

Carbonation is eventually inevitable, and is caused by the ingress of atmospheric carbon dioxide reacting with the pore water to form carbonic acid. This neutralises the alkalinity in the concrete. This occurs progressively, and a carbonation front moves through the concrete until it reaches the steel. The passive layer then breaks down as the pH falls from over 12 to around 8. Corrosion can start in the presence of oxygen and water as the pH falls below 11. The carbonation front moves approximately according to the following parabolic relationship:

Carbonation depth = Constant × Square root of time

A typical Portland cement concrete may show a carbonation depth of 5–8 mm after 10 years, rising to 10–15 mm after 50 years (BRE, 2000 (a)). Therefore structures with low concrete cover over the reinforcing steel will show carbonation-induced corrosion more quickly than those with good cover.

The rate of decay of reinforcing steel in concrete is also affected by the concrete quality. Concretes made with a high water to cementitious ratio and with a low cementitious content will carbonate more quickly than well-made concretes because they are more porous and have lower reserves of alkali to resist the neutralisation process. For concretes with pulverised fuel ash or other cement replacement materials, the drop of alkaline reserves is usually balanced by the increase in concrete quality for an equivalent Portland cement, except at high replacement levels in dry conditions (BRE, 2000 (a)).

The rate of carbonation is also affected by environmental conditions. Carbonation is more rapid in fairly dry and wet/dry cycling environments. It may therefore occur more rapidly in bathrooms and kitchens in blocks of flats, and in multistorey car parks where the carbon dioxide levels are high due to exhaust fumes.

Chloride attack

The second major cause of reinforcement corrosion is chloride attack. This is usually due to one of the following causes:

- de-icing salt ingress from roads and vehicles;
- sea-salt ingress in marine environments;
- cast-in salt from contaminated mix components;
- cast-in calcium chloride as a set accelerator.

A certain level of chloride is needed for corrosion to occur. Figure 4.2 illustrates the relationship between chloride level, carbonation and corrosion risk.

Once the chloride level at the reinforcement exceeds 0.4% by weight of cement (approximately 0.06% or 600 p.p.m. by weight of sample

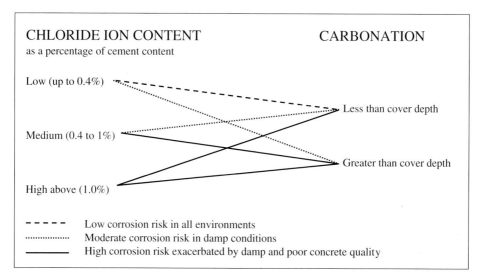

Fig. 4.2 Risk of corrosion in relation to concrete analysis (reproduced from *BRE Digest* 264, 1982).

assuming 15% cement content), there is a significant risk of corrosion, especially in the presence of moisture. Obviously, if cast-in chlorides exceed 0.4%, then the corrosion risk rises (see BRE, 2000 (a)).

For chlorides that are diffusing into the concrete (as a result of salty sea spray or de-icing salts, for example), it can be a useful approximation to calculate the progress of a 0.4% chloride by weight cement–corrosion threshold progressing through the concrete at a parabolic rate comparable to the carbonation rate equation above.

$$\text{Chloride threshold progress rate} = \text{Constant} \times \text{Square root of time}$$

The use of this approximation, along with sampling and interpretation, is discussed in the section on sampling below.

As with carbonation, chloride-ingress rates are a function of concrete quality and environment. For concrete with low reinforcement cover, and particularly for poor-quality concrete, chlorides can be transported rapidly by wetting and drying absorption and by capillary action that almost sucks the chloride-laden water into the concrete. The water then evaporates and leaves the salt behind. In good-quality concrete with good cover to the reinforcement, diffusion processes predominate.

The corrosion process

Regardless of the cause of corrosion, once the steel's passive-layer protection is lost, corrosion proceeds by the mechanism illustrated in Fig. 4.3. Corrosion of steel in concrete is an electrochemical reaction in which the major constituent of steel (iron) goes into solution as iron ions with a flow of electrons (electrical flow). This site is called the anode. Electrons are produced in this self-sustaining process and flow through the reinforcement towards cathodic sites where they react with oxygen and water from outside to produce additional hydroxyl ions. This is known as the cathodic reaction. As can be seen in the diagram, the cathodic reaction requires water and oxygen. The initial anodic reaction does not require anything, once the steel is depassivated, it is only when the iron has reacted to become the soluble ferrous ion that it will react with the hydroxide ion (the alkalinity in the concrete), and then with oxygen and water to create the solid rust whose volume increase will crack and spall the concrete.

The fact that oxygen is not required at the anode is important because the exclusion of oxygen from anodic areas without stifling the cathodic reaction will lead to dissolution of the reinforcement rather than cracking

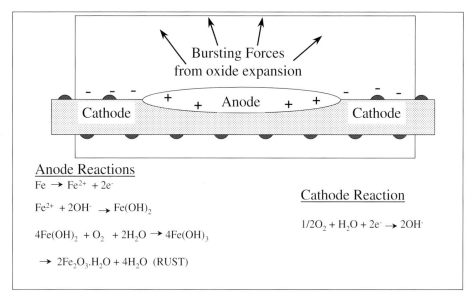

Fig. 4.3   The corrosion process for steel in concrete.

and spalling of the concrete. This can happen in local saturation conditions where the concrete is very wet and therefore conductive enough to allow good separation between anodes and cathodes.

The ingredients for corrosion are therefore:

- carbonation or sufficient chloride at reinforcement depth to depassivate the steel;
- oxygen to fuel the cathodic reaction and to create the expansive oxide;
- water to fuel the cathodic reaction and to create the expansive oxide;
- concrete of low enough resistivity to allow the electrochemical anode and cathode reactions to proceed.

These ingredients, along with the electrical nature of the reactions, can therefore be used to assess the corrosion condition, as described in the section on condition survey techniques below.

Design and construction defects

The performance of reinforced concrete can be severely reduced by poor design and construction techniques. These may cause reinforcement corrosion or degradation of the concrete itself, which in turn may lead to reinforcement corrosion.

Poor cover to the reinforcement

Insufficient concrete cover to the reinforcement is a major influence on the durability of reinforced-concrete buildings and structures (Fig. 4.4).

A number of problems, particularly with older structures, occur at the design stage, such as:

- older codes do not specify adequate cover, especially in saline environments;
- designers used to specify cover to the main steel, which meant that there was inadequate cover to stirrups, clips and so on;
- details such as drips, grooving of surfaces, and so on, reduced overall cover, often to vulnerable steel at corners and in areas of water run-off.

During construction a number of problems may arise, including:

- detailing that makes it difficult to achieve the specified cover;
- incorrect reinforcement placing;
- movement of reinforcement within shutters from the specified cover.

Fig. 4.4 Insufficient cover to reinforcement at Parkhill flats, Sheffield, is one of the most common causes of premature corrosion-induced concrete decay (photo Susan Macdonald).

Poor consolidation leading to honeycombing and voids

Poor consolidation of the concrete during pouring leaves honeycombing and voids in the concrete. These are frequently visible at corners and slab ends, where the cement paste does not completely surround the larger aggregate particles. Problems can occur over congested steel, where it is difficult to ensure that the concrete flows into all the interstitial spaces.

Concrete constituents and manufacture

There are a number of problems that can occur with the concrete mix at either the design or the execution stage. These are discussed in detail later in this chapter, and include:

- high water to cementitious products ratio—this leads to a porous concrete susceptible to carbonation and chloride ingress;
- cast-in chlorides—until the mid-1980s, calcium chloride set accelerator was considered to be a useful and safe admixture, as chlorides were considered to be bound into the concrete; similarly, seawater was used during construction on marine projects, a practice we now know to be problematic. Accidental inclusion of chlorides also occurs due to poorly washed mix constituents and contaminated water supplies;
- inadequate slump—the reverse of high water to cementitious product ratio, with inadequate superplasticiser giving rise to poor consolidation resulting in honeycombing and voids;
- alkali–aggregate reaction—where the aggregate reacts to the alkalinity of the cement.

Creep

Creep is caused by water being squeezed from the pores of the concrete owing to the sustained load of the concrete itself. Deformation due to creep is time-dependent, and it may not become apparent until some months after a structure has been completed. Allowance for deflection due to creep is made at the design stage. Where inadequate allowance has been made for creep, finishes may be affected by cracking. Openings may deflect, causing doors or windows to jam, for example.

Concrete degradation

Concrete is a relatively durable material, but it can be severely weakened by poor manufacture, the inclusion of damaging constituents or aggressive environments. As mentioned above, there are a number of causes of degradation of the concrete matrix.

Alkali–aggregate reactivity

As described previously, concrete pore water is highly alkaline. Unfortunately, some aggregates used to make concrete react with the alkalinity to form products that swell and damage the concrete.

The most common alkali–aggregate reaction is an alkali–silica reaction, known as ASR. Soluble silicates in the aggregates react and form silica gels. These gels absorb water and expand. The result is a 'map cracking' effect and efflorescence of the gel, as shown in Fig. 4.5.

Many aggregates exhibit ASR to a greater or lesser extent when examined by petrographic analysis under a microscope. A more limited number show serious problems which are now well characterised in terms of type and source. Frequently, ASR will occur on a structure or part of a structure, the susceptible aggregates will react and then the situation will stabilise. The problem is frequently one of appearance rather than anything else. In a series of over 100 structures with ASR in the UK, typically 0.03 mm/m/year to 0.3 mm/m/year expansion occurred over 20 years (Wood, 1990). It may be possible to slow ASR by

Fig. 4.5 Map cracking due to an alkali–silica reaction.

reducing or eliminating moisture either by deflecting the run-down, or by the application of coatings or sealers.

Sulfate attack

Sulfates of sodium, calcium, potassium and aluminium are found in groundwater and soils and can cause degradation of the concrete matrix by expansive attack of the calcium hydroxide and calcium aluminates in the concrete. Wet/dry cycling causes salts to be accumulated on the concrete surface, resulting in degradation. Delayed ettringite formation, and the recently discovered problem of thurmasite attack of concrete, are also forms of sulphate attack. More than 0.1% water-soluble sulfate in soil or 150 p.p.m. in water is moderate exposure to sulfate attack. More than 2% in water or 10000 p.p.m. in soil is severe exposure. A high-quality dense, blended cement concrete gives good sulfate resistance. Low C_3A content cements have the disadvantage of low resistance to chloride ingress, so in cases where the sulfates are from marine exposure, sulfate-resistant Portland cement (ASTM Type II) may accelerate chloride-induced corrosion.

High-alumina cement concretes

High-alumina cement (HAC) was used extensively in the 1960s and 1970s to achieve very high early strength concrete. HAC also has a higher level of resistance to acids and sulfates. Under certain conditions during its curing (high water:cement ratio and high temperatures during curing) and particular environmental conditions after construction (high temperatures and/or high humidity levels), it suffers from a chemical (crystal) change and severe loss of strength and porosity. The cement is then attacked by some chemicals, such as calcium sulfate found in gypsum plasters. This has resulted in a number of serious structural failures. A major programme of identification of HAC structures has been undertaken in the UK and most other countries where it was used; HAC beams and columns have been replaced, supported or are carefully monitored. HAC also has very low reserves of alkali, so it carbonates very easily.

Environmental influences

In addition to the damage caused to concrete as a result of exposure to environmental agents such as carbon dioxide and acid gases, there are a number of other environmental factors that can cause deterioration.

Weathering

Staining

Water-staining of unpainted concrete is a major problem in the UK. The porosity and water-absorption characteristics of concrete seem to make it more susceptible than brick and stone to this type of soiling. Concrete can be cleaned like any natural stone or brickwork, and similarly removing surface laitance can make it more susceptible to future staining. Cleaning is discussed in detail in Chapter 6.

Erosion

Continuous water run-down, especially of water containing suspended solids, will erode concrete with time. Salt water and other aggressive solutions will etch concrete.

Efflorescence

Efflorescence can occur due to ASR, water run-off, or water running through slabs and carrying away the soluble calcium. Figure 4.6 shows an extreme case of a car park with stalactites and stalagmites forming where water runs through an unmaintained suspended slab.

Freeze–thaw damage

Freeze–thaw damage occurs where water accumulates in cracks, voids or pores of the concrete, and then freezes quickly enough so that the expansion cracks the concrete. Air-entrainment produces pores and small voids in the concrete of a size to avoid freeze–thaw. Older structures (that may not have been well air-entrained) and those in a severe heating/cooling environment will suffer from the scaling away of the concrete surface due to freeze–thaw damage.

Structural damage

Structural damage can result from;

- inadequate design or construction;
- settlement or other ground movement;
- overloading or change of use;
- fire damage;

Fig. 4.6(a)   Poor maintenance of this car park has lead to the formation of a stalagmite where the inadequate drainage keeps water off cars but does not cure or control leakage from the floor above.

- impact damage;
- seismic damage;
- wind damage.

These are usually easily identified by a qualified engineer. There are well-proven methods to rectify these problems. The importance of a structural appraisal in determining a concrete building's suitability for use is discussed in Chapter 3.

In most Western countries, there are usually sufficient resources and expertise to prevent corrosion from becoming a structural problem. However, the removal of the cover concrete by the expansive growth of oxides will reduce the bond between steel and concrete, which could

Fig. 4.6(b) In this photograph from the same car park stalactites have formed around a leaking drain.

allow bars to bend or buckle, or allow stirrups to corrode through. Any of these factors may lead to structural damage.

The greatest structural corrosion risk is on prestressed concrete structures. If post-tensioning tendons are inadequately protected against corrosion, then there is a risk of catastrophic failure as they are typically loaded to 80% of their load-bearing capacity. Tendons in conduits in prestressed concrete, for example, are very difficult to examine by normal non-destructive methods, so failures can go undetected. In North America, there are apocryphal tales of anchorages flying out of buildings as tendons fail on post-tensioned floor slabs.

Identification of defects, damage and decay mechanisms

The importance of correctly identifying the causes of concrete damage, defects and decay, and the need and methodology for carrying out appraisals and assessments, has already been discussed in Chapter 3. The following section describes the various tools and methods available for carrying out a detailed investigation in order to assess the condition of concrete buildings and structures.

There are a huge number of techniques for identifying and measuring defects and damage in concrete buildings and structures. The most important is the human eye and experience. A good engineer or technician, experienced in condition surveys, will know the differences between cracks produced by corrosion, ASR and thermal or other movement. This alleviates the necessity for extensive non-destructive testing to identify the source of the problem. Correct identification of defects and decay mechanisms is imperative to ensure that the repair is correctly targeted to the problem.

It is essential to plan any investigation carefully regardless of its scope, to ensure that any latent damage is revealed. Any investigation is reliant on the skills of the investigator, and their ability to utilse the tools and to interpret the results. Careful planning will need to take into account the affects of noise, the disruption to any building users and access issues. There may be aesthetic issues to consider and if the building is listed, statutory consent may be required prior to any testing or sampling procedures.

The principal aim of the investigation is always thoroughly to evaluate the condition and performance of the building or structure; that is, identify the causes, measure the extent and progress of the damage, and assess this against necessary performance. This information is used as the basis for the repair strategy.

Condition survey techniques

Non-destructive testing techniques are usually required to supplement and confirm the engineer's diagnosis, and to quantify the extent of deterioration for repair measurements. No one technique will provide all the answers, especially if the problem is reinforcement corrosion. The most common techniques are described briefly in the following section.

Visual inspection

Visual inspection is the vital first step in any investigation. A site survey should include drawings or a schematic diagram of the structure marked with the visual defects. These should be taken from an extensive key of defects, as illustrated on the structure in Fig. 4.7. Photographs and videos can also play an important role in recording the results of a visual survey, but the locations must be recorded accurately.

Hammer or delamination survey

If corrosion is suspected, then a hammer is used to sound the concrete and identify hollow areas where corrosion has led to cracking between

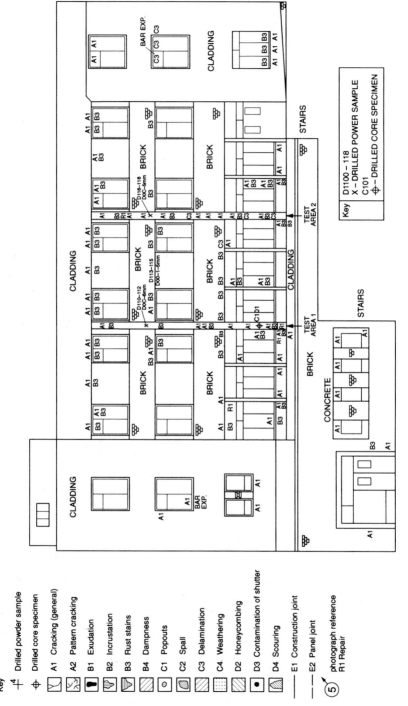

Fig. 4.7 Typical visual survey of a reinforced concrete building. (From Broomfield (1997) courtesy of E & FN Spon.)

bars, with negligible visual signs of damage. This is usually marked on the visual survey drawings.

Cover survey

A magnetic cover meter is used to check the concrete cover over the steel. It can also be used to check the diameters of the steel bars, although the results should always be confirmed from original reinforcement drawings and from selectively breaking out bars.

Depth of carbonation

Phenolphthalein indicator is used to measure the depth of carbonation, i.e. the location of the carbonation front and its proximity to the bars. Here, the solution is sprayed onto freshly broken concrete, which turns pink where the concrete is still alkaline. The method of making up the solution and applying it is described in BRE (2000 (a)). The test can be done on-site or in a test laboratory on cores taken from the site. Obviously, as the average carbonation depth approaches the average cover depth, more steel will start to corrode.

Chloride analysis

It is standard practice to take representative drillings of concrete and take them to the laboratory for chloride content analysis. Drillings can be taken to different depths, e.g. in 10-mm increments, to determine the profile of the chloride in the concrete. As the chloride content rises at reinforcement cover depth, so the likelihood and extent of corrosion increases.

In the laboratory, drillings are usually dissolved in acid and then titrated to measure the chloride content, or analysed using a chloride-specific ion electrode. There are site procedures using 'Quantab strips' and field chloride ion electrodes, but it is usually more economic and accurate to use a quality-assured in-house test for analysis.

There has been some discussion in the technical literature about the merits of measuring the water-soluble chloride content rather than the acid-soluble chloride. The concept being that as some of the chloride is bound up in the cement paste as chloroaluminates, only the water-soluble chlorides contribute to corrosion. However, it is difficult to measure water-soluble chlorides, as the test techniques bring out some of the bound chlorides to different extents depending on the details of preparation (how small the sample is ground, how long it is soaked in water, what temperature it is tested at). Therefore, the usual practice is to stick to measuring the acid-soluble chloride content and use 0.4% and 1.0% chloride by weight of cement thresholds for corrosion, as shown in Fig. 4.2.

Sampling

Sampling for petrographic analysis usually requires a core, normally about 100 mm in diameter. Smaller cores can be used as long as the diameter is at least twice the maximum aggregate size. Petrography can be used to examine for the silica gel formation indicative of ASR, and to look at consolidation, pore sizes and distribution, freeze–thaw damage, cement content, aggregate types and any other suspected problems.

Cores are also used to measure the concrete resistivity as long as the moisture is sealed in on removal and the core is not excessively wetted during the coring procedure. This process is described further in the section on resistivity measurement.

Core samples can also be sliced and crushed for chloride analysis to provide a profile with depth. This can be used to predict the rate of chloride build-up at reinforcement depth.

It is important that core samples are representative of the area of interest. Coring can affect the appearance of the building, and core sites need to be carefully selected to ensure that they are useful, but that the visual impact is minimised. Listed buildings may require statutory consent prior to coring.

Radar, thermography, radiography and ultrasonics

These techniques are used to look into the concrete non-destructively. They can find cracks, voids, prestressing tendons, reinforcement and other embedments. They require specialists to apply and interpret them.

Endoscopy

Endoscopes (fibre-optic viewers) can be inserted into voids or ducts in the concrete to view the internal condition of the concrete. They can be especially useful for examining post-tensioning ducts for internal corrosion.

Resistivity measurement

Resistivity is an indicator of concrete quality and moisture content. A high-resistance concrete will resist corrosion, as the ions cannot circulate between the anode and the cathode. Resistivity can be measured with a four-probe Wenner array on-site, or can be taken from a core, as described above. A laboratory core can be conditioned in a constant-humidity environment to compare its resistivity with other concretes. Resistivity can be used to gauge corrosion susceptibility (Broomfield, 1997).

Half-cell potential survey

As corrosion is an electrochemical phenomenon, the location and magnitude of the anodes and cathodes can be plotted with a simple device called a half-cell. In simple terms, a half-cell is a metal rod in a solution of its own ions, for example silver in silver chloride, with a semi-porous membrane for ionic contact (or porous plug). If the half-cell porous plug is placed onto the concrete surface it is in ionic contact with another half-cell, i.e. iron in a solution of its own ions Fe^{2+} in the form of ferrous hydroxide $Fe(OH)_2$, as shown in Fig. 4.3. This forms a simple electrical cell.

By connecting a voltmeter between the reinforcing steel and the half-cell, a voltage develops. As the half-cell is moved across the concrete surface, the voltage changes as the concentration of ferrous ions changes. Grid readings can be taken and the data can be plotted on a plan of the structure to produce contour plots of equal potential showing the most anodic areas (ASTM, 1991; Chess and Gronvold, 1996; Broomfield, 1997, p. 40–48). An example is shown in Fig. 4.8.

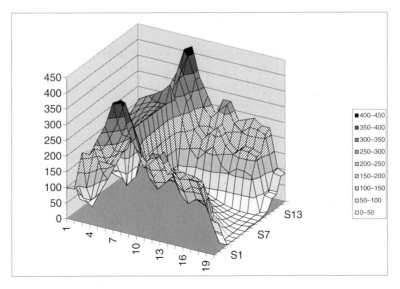

Fig. 4.8   Half-cell potential survey.

The higher the magnitude of the potential, the more active the anode. However, this technique only shows the more anodic areas and indicates the corrosion risk; it is not a measure of corrosion rate, which requires more sophisticated equipment, as described in the following section. Readings can be misleading in circumstances of water saturation, carbonation and high-electrical-resistance concrete.

Corrosion rate measurement

There are two methods of measuring the corrosion rate of steel in concrete. Both of them measure the current flow between anodes and cathodes indirectly. The most widely used technique is the linear polarisation, or polarisation resistance, technique. This is a half-cell with an auxiliary electrode attached. After the half-cell potential is measured, a small current is applied from the auxiliary electrode. This shifts the half-cell potential. The corrosion rate is determined as follows:

Corrosion rate $=$ (Constant $\times$ Applied current)/Potential shift

It takes a minute or two to take a corrosion-rate reading, and it must be located over a known area of rebar, so the process is far slower than a potential survey. Corrosion-rate measurements are generally taken at areas identified as interesting during a visual, hammer and half-cell survey. They are frequently used to help to model the deterioration rate.

Galvanic current measurement

The alternative to the polarisation resistance technique is to measure an actual current flow. This is usually done by introducing artificial anodes and cathodes that are electrically isolated and so can be connected through an ammeter to measure the current. In some cases it may be possible to isolate short sections of reinforcing steel in anodic areas, and measure the current either from the steel section to the rest of the steel, or to an artificial cathode, usually made of stainless steel, embedded into the concrete. It is important to avoid disturbing the steel more than necessary so that it is in the same environment as the rest of the reinforcement network. Galvanic or macro-cell probes are more usually incorporated during construction.

Corrosion monitoring

As an alternative, or in addition to a snap-shot survey of a structure, some probes can be permanently embedded to monitor changes in the condition of the concrete and reinforcement. These include half-cells, and corrosion-rate and resistivity probes. They can be linked to modems and hence to a remote computer, or to loggers or a hand-held unit brought to the site at regular intervals, typically between one and four times a year, to download data or to take measurements at locations of interest. Monitoring may be incorporated in a long-term repair/maintenance strategy as an alternative approach to a one-off repair. It can be useful where treatments are carried out, especially where coatings, barriers or cladding are installed and the surface can no longer be accessed for surface-applied NDT techniques. It can also be useful where more experimental treatments such as surface-applied corrosion inhibitors are used, or where treatments have indeterminate lives, such as electrochemical realkalisation or chloride extraction, as discussed in the next chapter.

Defect monitoring

Documentary evidence of the building's condition in the past, such as photographs and previous reports, are useful ways of measuring the progress of deterioration. This is an important part of any appraisal process, as discussed in Chapter 3, and can also be used to assess the timing (urgency and prioritisation) of proposed repair works. It is therefore important that the visual survey is as quantitative as possible in recording the location and size of defects so that their growth (if any) can be monitored.

Structural monitoring, environmental monitoring and corrosion monitoring are all tools in the assessment process. A number of references on this topic are provided in the further reading section.

References

ASTM (1991) *Standard Test Method for Half-Cell Potentials of Uncoated Reinforcing Steel in Concrete*. ASTM C876-91. ASTM, West Conshohocken, PA.

BRE (2000) (a) *Digest 444. Corrosion of Steel in Concrete. Part 1. Durability of Reinforced Concrete Structures. Part 2. Investigation and Assessment. Part 3. Protection and Remediation*. (b) *Digest 330. Alkali–Silica Reaction in Concrete. Part 1. Background to the Guidance Notes. Part 2. Detailed Guidance for New Construction. Part 3. Worked Examples. Part 4. Simplified Guidance for New Construction Using Normally Reactive Aggregates*. Building Research Establishment, Garston.

Broomfield, J.P. (1997) *Corrosion of Steel in Concrete: Understanding, Investigation and Repair*. E & FN Spon, London.

Chess, P. and Gronvold, F. (1996) *Corrosion Investigation: A Guide to Half-Cell Mapping*. Thomas Telford, London.

Wood, J.G.M. (1990) Physical effects of AAR: structures as a laboratory. In: Soles, J.A. (ed) *CANMET International Workshop on Alkali Aggregate Reaction*. CANMET, Halifax, Canada.

Further reading

ACI-364 (1993) *Committee 364. Guide for Evaluation of Concrete Structures Prior to Rehabilitation*. ACI 464.1R–94. American Concrete Institute, Detroit.

BRE (1991) *Digest 366. Structural Appraisal of Existing Buildings for Change of Use*. Building Research Establishment, Garston.

BSI 1881 (1986) *Testing Concrete: Guide to the Use of Non-Destructive Methods of Test for Hardened Concrete. Part 201*. British Standards Institution, London.

Concrete Society (1992) *Non-Structural Cracks in Concrete. Technical Report No. 22*, 3rd edn. Concrete Society, Slough.

Neville, A.M. (1981) *Properties of Concrete*, 3rd edn. Pitman, London.

Pullar-Strecker, P. (1987) *Corrosion-Damaged Concrete: Assessment and Repair*. Butterworth, London.

Standards Australia, Standards of New Zealand, ACRA and CSIRO (1996) *Guide to Concrete Repair and Protection*. Sydney.

Watt, D. (1999) *Building Pathology: Principles and Practice*. Blackwell Science, Oxford.

Watt, D. and Swallow, P. (1996) *Surveying Historic Buildings*. Donhead, Shaftesbury.

Chapter 5
Repairing Damaged Concrete
John Broomfield & Susan Macdonald

Introduction

The previous chapters have reviewed the various concrete deterioration processes, investigative methods and their assessment. Once the cause and extent of deterioration has been determined and an appraisal has assessed the structure's fitness for purpose, then the type of repair and any other remediation work can be considered. There is a wide range of issues that determine the type of repair conducted, many of which are building- or structure-specific. The information provided here is therefore general in nature, and reviews the strengths and weaknesses of different repair techniques.

There are a number of factors that affect the repair and remediation strategy that will be appropriate to a specific structure; these are summarised in Fig. 5.1.

Repair options

There are various strategies that can be adopted to address the defects, damage and decay of a concrete building or structure, and these include (DD ENV 1504-9, 1997, Part 9):

- do nothing;
- re-consideration of the building's structural capacity—possible downgrading of function so it is fit for purpose;
- arrest or reduce further deterioration;
- repair, remediation and any necessary structural upgrading;
- reconstruction of part or all of the structure;
- demolition.

As discussed in Chapter 3, the important first step is a review of the available data, including the original construction drawings as well as the

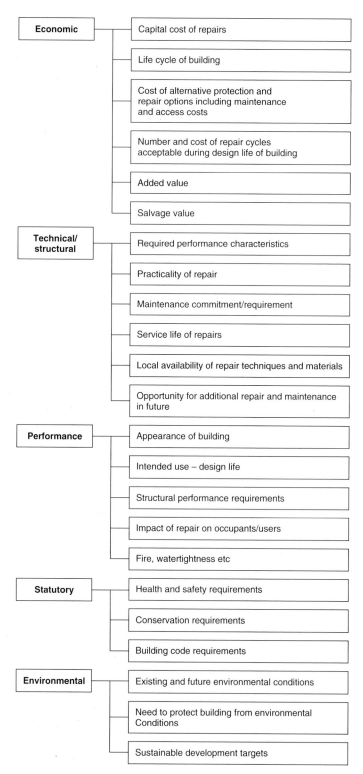

Fig. 5.1 Factors affecting the repair and remediation strategy (Susan Macdonald).

condition survey data. This may lead immediately to the selection of the only available repair or remediation option, or it may eliminate some options.

A significant issue is the required residual service life of the structure. If it only has a residual service life of a few years (typically less than ten), then short-term options such as patching and sealing, or enclosing and possibly supporting damaged areas, may be acceptable. If the structure is important and needs to be retained (such as a listed building), it may warrant more complex and expensive repairs. Most structures will fall in between these two extremes.

If budgets are limited then repairs may need to be staged, or more expensive options eliminated. Occupied buildings may have scheduling issues, or requirements to minimise noisy break-out and drilling of concrete. Night and weekend work may be necessary on office buildings; daytime work will be preferred on dwellings. The ability to carry out the maintenance required as part of the long-term strategy will also affect the repair selection.

When selecting a treatment for damaged concrete, the advantages and limitations of the different treatment methods must be evaluated. Repair techniques range from doing nothing to more complex treatments or combinations of treatments. DD ENV 1504-9 (1997) (currently in draft for comment for the specification of concrete repair works, but likely to become mandatory in Europe in the near future) categorises repair systems under 11 different principles.

1. Protection against the ingress of agents (such as water, gases, chemical vapour, other liquids and biological agents) by impregnation, surface coating, crack bandaging, filling or transferral to joints, and membrane application.
2. Moisture control within a specified range of values by impregnation, surface coating, sheltering or overcladding, or electrochemical treatment.
3. Concrete restoration by mortar application/patch repairs, recasting, sprayed overlays and replacement.
4. Structural strengthening to increase or restore load-bearing capacity by the addition or replacement of rebars, bonded rebars, additional concrete, crack/void injection and prestressing.
5. Increasing physical resistance to physical or mechanical attack by overlays or impregnation.
6. Increasing chemical resistance by impregnation, overlays or coatings.

7. Preserving or restoring the passivity of the reinforcement by increasing concrete cover, replacing contaminated or carbonated concrete, realkalisation or electrochemical chloride extraction.
8. Increasing the resistivity of the concrete by limiting moisture using surface treatments, coatings or sheltering.
9. Cathodic control to prevent anodic reaction by surface coating or saturation.
10. Cathodic protection.
11. Anodic control to prevent anodic reaction by painting the reinforcement with active pigments or barrier coatings and inhibitors.

Depending on the cause of the problem and the required outcome, one or more of the repair options discussed in this chapter may be adopted as part of the repair and remediation strategy. A structure may be suffering from various different types of problems with different causes, and therefore a combination of repair methods may be necessary to address each problem. Each treatment has its technical advantages and limitations, and must be used appropriately. The choice of which treatment to use, either alone, in combination or in series, is determined by its technical merits, its appropriateness for the structure, and by life-cycle cost analysis. Once corrosion is initiated, the engineer is unlikely to find a 'zero maintenance' option. A monitoring or maintenance cycle will be required on almost all structures, except those with very short residual service lives. Therefore, any repair strategy should be designed in association with a long-term maintenance program. It is difficult to separate repair and maintenance strategies. This chapter deals with methods limited to the repair of existing concrete; techniques that are part of longer-term repair, prevention and maintenance strategies are discussed in more detail in Chapter 6.

Do nothing

It may be acceptable to do nothing if either a short residual service life is acceptable or it is possible to await full or partial replacement. If there is a risk of concrete falling off the structure, this may be supplemented by enclosure, or some form of containment. Corrosion monitoring may be useful to determine the rate and extent of deterioration.

Concrete restoration systems

These include patch repair, recasting, sprayed overlays and replacement of concrete.

Patch repairs

Patch repairs are usually required prior to any other treatment, wherever there is visible damage. Depending on the cause of deterioration and the future requirements, properly executed patch repairs can provide a very effective long-term solution to deterioration.

The Concrete Society has produced two technical reports on the repair of reinforced concrete (Concrete Society Technical Reports No. 26 (1985) and No. 38 (1991)). Both reports emphasise the importance of carrying out good-quality repairs with suitable materials.

Patches can be hand-applied, poured or pumped into shuttered repairs, or spray-applied. In each case the repair material must be carefully chosen and applied by trained and experienced applicators. Tests on bond strength, compressive strength and consolidation should be carried out as part of a properly specified repair contract.

Proprietary patch repair materials may be more expensive than site-batched sand/cement mixtures, but they ensure consistent good-quality, low-shrinkage repairs that bond well to the parent concrete. However, for structures where the concrete finish is an architectural feature, proprietary repair mixes may not provide the best visual match. Either way it is difficult to match new concrete to the parent concrete cosmetically, and if they do match when new, they will usually weather differently so they will show up later (Fig. 5.2). Coating may therefore be required for aesthetic reasons. This clearly poses a dilemma where the exposed concrete finish is an important part of the structure's aesthetic. It is essential to prepare the cut-out area properly with no feathered edges, cutting behind the rebar far enough along to minimise recurrent problems (less essential when using electrochemical techniques, particularly cathodic protection), and to compact the repair properly.

As with the deterioration of any building fabric, it is imperative to diagnose the cause of concrete deterioration correctly in order to identify the appropriate repair method. Patch repairs are not an appropriate long-term method of preventing deterioration where the structure contains a high chloride concentration. The repair of chloride-contaminated concrete may involve patch repair of the visually damaged concrete. The appropriate repair methods may be similar to those proposed for structures suffering from other causes of decay, but the long-term performance of the treatment and how it is applied may be different for chloride-contaminated concrete (Concrete Society technical reports No. 26 (1985) and No. 38 (1991)).

The reasons for this are illustrated in Fig. 5.3. As discussed in the previous chapter, corrosion of steel in concrete occurs by the formation of anodes and cathodes. The formation of an anode may force cathodes to

Fig. 5.2 Patch repairs at the Alexandra Road Housing Estate in Camden, London. This is a Grade II* listed building, and therefore it was important to match the repairs as closely as possible to the parent concrete. The high-quality, board-marked concrete finish was an important feature of the architect's design. A latex mould was used to replicate the timber-board-marked finish (photo Susan Macdonald).

form around it, even though the chloride level may be above the corrosion threshold at the cathodic reaction sites. The anode provides galvanic cathodic protection to the area around it. Once a patch repair is carried out, the cathodes no longer have an anodic reaction to sustain them, and may become anodes in their own right and therefore corrosion can occur.

Patch repairs will not necessarily work structurally to accept load, so a proper structural analysis is required to ensure that the structure is still safe (see Chapter 3). Care must also be taken to retain structural integrity

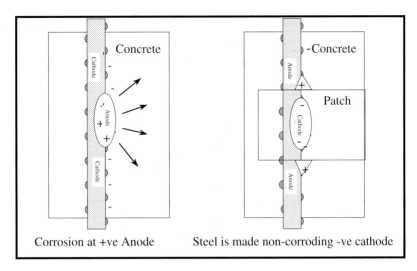

Fig. 5.3 The formation of incipient anodes after patch repairing (John Broomfield).

when cutting out concrete behind the reinforcement bar to remove carbonated or chloride-contaminated concrete. Structural propping may be required, or the load may have to be reduced on the element being repaired. Patch repairs are often carried out in combination with other repair methods.

Concrete replacement

Where larger areas of damage have occurred, it may be feasible to remove all corrosion-damaged and chloride-contaminated concrete, or to replace individual elements. This has specific logistical and financial implications that will have to be assessed when considering the different repair options.

Sprayed overlays

A wide range of sprayed overlays are available for a variety of applications. They can be used in concrete repair to rebuild the profile in large areas of removal, in strengthening works and as overlays in cathodic protection systems. The highest level of workmanship is required to produce durable, good-quality sprayed concrete, with good surface preparation that will ensure long-term adhesion to the substrate.

Barrier and impregnation systems

Barrier and impregnation systems are used to reduce or prevent the ingress of chemical or environmental agents, to control the moisture in the concrete and to provide resistance to chemical or physical attack.

Coatings, sealants and membranes

Paint coatings, penetrating sealers or membranes may be applied to elements of concrete structures exposed to severe conditions, such as salt ingress, at the time of construction or as part of a repair treatment.

In the UK it has not been usual practice to coat concrete at the time of construction, although waterproofing membranes are put on bridge decks and some car park decks to keep out chlorides from de-icing salts. Penetrating sealers such as silanes, or more specifically siloxy silanes, are now used on bridge substructures to stop salt spray and run-down ingress.

If applied as part of a repair, coatings can be effective in reducing chloride ingress or carbonation progress (Figs. 5.4 and 5.5). However, if sufficient chlorides are already present at rebar depth to initiate corrosion, then membranes and coatings will not prevent it. Penetrating sealers have been shown to be effective in drying out concrete, and this will reduce the corrosion rate (see previous chapter). It is important to use the right coating for the right job. Penetrating sealers will not stop carbonation, and may accelerate it. Anti-carbonation coatings should have test certificates from independent test houses which show their effectiveness. As barrier systems play an important role in maintenance strategies, these are discussed in more detail in Chapter 6.

Corrosion inhibitors

Diffusing corrosion inhibitors into the concrete to protect the steel is a new approach to concrete repair. The use of calcium nitrite as a corrosion-inhibiting admixture in the concrete mix is well established. However, trials of inhibitor treatments to hardened concrete after corrosion damage has been observed are comparatively recent.

The present range of inhibitors can be summarised as follows:

- several proprietary formulations of vapour-phase inhibitors, based on volatile amino alcohols that create a molecular layer on the steel to stop corrosion;

Fig. 5.4 Pullman Court, Streatham Hill, London, designed by Sir Frederick Gibberd, 1936. The concrete was repaired by patching, followed by the application of an anti-carbonation coating to reduce the progress of carbonation (photo Susan Macdonald).

- calcium nitrite (an anodic inhibitor) in a mixture to aid penetration into concrete;
- monofluorophosphate, which seems to create a very alkaline environment as it hydrolyses in the concrete.

However, new materials are being produced and tested as concrete materials manufacturers see the potential market for inhibitors.

In principle, corrosion inhibitors are applicable in any situation. However, due to the present understanding of the materials available and the limited field-testing done to date, the authors would consider inhibitors when:

- the concrete has carbonation or low-to-modest chloride levels (less than 1% chloride by weight of cement);
- there is low cover to the reinforcement (less than 20 mm);
- the concrete is penetrable (carbonated or corrosion-damaged in less than 20 years);

Fig. 5.5 Patch repairs in progress at Pullman Court. Note the thinness of the walls, which is characteristic of early modern architects' work in Britain. The concrete has always been painted (Susan Macdonald).

- a barrier coating is applied after application;
- corrosion monitoring is installed to assess the effectiveness over time.

The main advantage of corrosion inhibitors is that application is comparatively inexpensive and simple for any element with an accessible surface. However, the knowledge and understanding of the range of inhibitors is limited at present. John Broomfield has conducted two field trials, one large and extensive that was not very successful, and one very small and short-term which gave encouraging results. The issues that need consideration in relation to inhibitors are:

- do they penetrate the concrete?
- can they spread uniformly and effectively along the rebar network?
- can they suppress corrosion?
- does corrosion stay suppressed?
- for how long are they effective?
- to what level of chloride or rate of corrosion are they effective?
- can we measure all these things?

All of these issues need independent investigation in the laboratory and in the field for a range of inhibitors.

Electrochemical techniques

Patch repairs treat damaged areas of concrete, and coatings, sealants and membranes slow down carbonation and chloride ingress rates and can keep concrete dry. However, when chloride is present in damaging concentrations, or when the depth of the carbonation has reached a critical level, electrochemical techniques can be used to provide overall protection to large areas of a structure for long periods of time.

These techniques use an artificial anode to pass a current to the reinforcing steel, which is then made a cathode. The treatment may involve the short-term passage of a current from a temporary anode, or a permanent system, as in the case of cathodic protection.

Although more complex and often of higher initial cost than conventional repairs, these techniques, and cathodic protection in particular, are frequently the most cost-effective options for structures with more than 10 years residual service life after treatment where chloride or carbonation damage is extensive. As the electrochemical treatment is stopping the corrosion, patch repairs prior to treatment are less extensive, being confined to areas of visible damage. The structure is less likely to require propping. For occupied buildings, the reduction in patching will reduce noise, dust and inconvenience to occupants.

Cathodic protection (CP)

There are two types of cathodic protection (CP) systems—impressed current or galvanic. The galvanic systems are related to the concept of galvanizing steel, but it is less well established when the steel is embedded in concrete. Impressed current CP requires a permanent mains power supply to run the system, and regular monitoring and adjustment by qualified personnel. If remote monitoring is installed, then a permanent telephone line is usually required.

Impressed current cathodic protection (ICCP)

Impressed current cathodic protection (ICCP) is now a well-established technique for the treatment of chloride-induced corrosion on buildings and other structures. The first trials in the UK were carried out by John Broomfield on government buildings in 1986. Somewhere between two

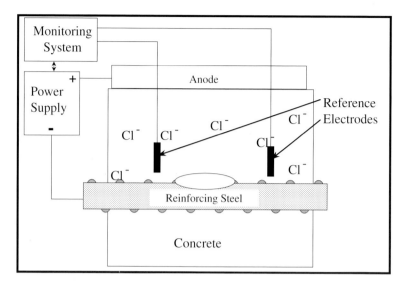

Fig. 5.6   Impressed current cathodic protection (John Broomfield).

million and three million square metres of cathodic protection anode material is installed on everything from steel-framed historic buildings to the Sydney Opera House walkway, bank vaults in the Middle East, and major wharf and bridge systems. Cathodic protection is used wherever long-term protection is required on corrosion-damaged structures with a significant residual life. A typical project covers two or three thousand square metres. However, many small systems of a few tens of square metres exist, particularly on historic and residential buildings and structures.

Figure 5.6 shows a typical ICCP system. It shows the essential parts of the system, which are:

- a permanent distributed anode system;
- a power supply which converts mains power to low-voltage d.c. (typically 10 V at 1 A);
- embedded reference electrodes to monitor the effectiveness of the system (see Chapter 4 on embedded half cells);
- a control system which can be manual or computer-driven, and either locally or remotely monitored;
- the cabling system to carry the power and signals, including connections to the reinforcing steel.

An ICCP system should be designed, specified and installed by suitably qualified and experienced personnel. Although the various elements

Table 5.1 Types of anode for impressed current cathodic protection.

Anode type	Application area	Relative weight	Approx. life (years)	Comments
Titanuim mesh in an overlay	All areas	Increase in dead weight	Up to 120	Durable, established system. Main problem is overlay application
Titanium ribbon in slots	All areas	Negligible	Up to 120	Needs sufficient cover to reinforcement
Conductive organic coatings (paints)	Away from water and wear	Negligible	Up to 15	Very popular, inexpensive anode system for buildings and substructures
Probe or discrete anodes	Most areas	Negligible	Up to 50	Fairly new. Needs careful design and installation, and checks on the number of holes needed
Conductive mortar	Most areas	Increase in dead weight	Up to 25	Fairly new, excellent adhesion. May need overlay
Thermal sprayed zinc	Most areas except trafficked	Negligible	Up to 25	Zn used in USA widely. More expensive than paints, but more durable, especially in wet conditions. Zinc can be used galvanically in some circumstances

of the cathodic protection use conventional expertise (concrete repair, anode application, wiring), proper design is essential to ensure that all areas receive the necessary protection, and that the most appropriate anode system is used. There are a number of different anode types, including conductive coatings (paints or thermal sprayed metals), overlay systems (coated titanium mesh in sprayed concrete or sprayed conductive mortar), or rods, tubes or ribbons embedded in holes or slots cut into the concrete. Table 5.1 describes the main anode systems currently in use and their applications. The table is not comprehensive, but indicates the relative merits and limitations of different anode systems. New anodes are continually being developed. Selection of the most appropriate system is dependent on the building or structure design, and the usual cost and logistical factors.

ICCP has a comparatively high initial cost and the anodes generally need replacing on a 10–40-year cycle, depending on the type used. However, life-cycle cost analysis frequently shows ICCP to be one of the most cost-effective repairs for structures with 20 years or more of residual service life. Once applied, the system stops the cycle of regular concrete repair associated with structures with high chloride levels.

If carefully designed, an ICCP system can provide a less-invasive method of concrete repair than traditional methods. For listed buildings, or where the aesthetics of the structure are important, this can be an added advantage.

ICCP requires continuous on-going monitoring for the life of the system/structure by suitably trained and qualified persons. The advantage of continuous monitoring is that the system's effectiveness and the condition of the structure are under continuous scrutiny, and therefore if further intervention is required it can be carried out in a timely manner.

There are some situations where ICCP may not be the most effective treatment. ICCP cannot easily be applied to elements containing pre-stressing steel. There is also concern about its application to structures with a high susceptibility to alkali–silica reaction (ASR). Structures with a lot of electrically discontinuous steel can be expensive to protect, as all the steel must be in contact for the current to flow correctly. This includes epoxy-coated reinforcing steel, where bars are electrically isolated by their coatings. Short circuits caused by tie wires and tramp steel touching the anode can also be a problem.

Current cannot flow through insulators, so coatings and membranes must be removed prior to an ICCP system being installed. ICCP anodes evolve gases, so if impermeable membranes are being applied over the anode then gas evolution must be dealt with.

Additional sources that provide more comprehensive information on ICCP and its background, theory and technology are provided in the further reading section at the end of this chapter.

Galvanic/sacrificial anode cathodic protection

Figure 5.7 illustrates a galvanic or sacrificial anode cathodic protection system. This is far simpler than an ICCP system, as the anode corrodes to provide the current to protect the steel. However, since concrete has a high resistance, galvanic systems are usually applied to structures in very wet conditions, such as tidal marine exposure. Recent developments are leading to trials of humectants, which humidify the concrete and lower the resistance, making galvanic cathodic protection more viable in a wider range of applications. At the moment this is used with the thermal sprayed zinc anode described in Table 5.1. More recently a zinc-foil, glue-on anode has been developed with a conductive hydrogel adhesive to attach it to the concrete. This anode leaves the structure with a sheet-metal finish that dulls with time, or which can be painted.

A recent innovation is a small (approximately 100 mm diameter) plug anode that can be wired to the reinforcing steel in a patch repair. This will avoid the incipient anode effect described in Fig. 5.3. This offers an option for improving simple patch repairs to those unwilling or unable to install ICCP.

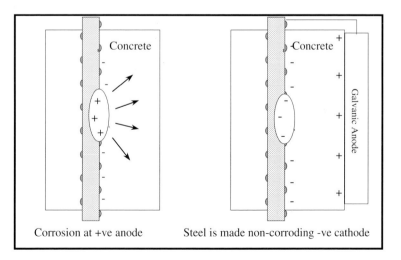

Fig. 5.7 Galvanic cathodic protection with sacrificial anode to stop corrosion (John Broomfield).

Electrochemical chloride extraction (ECE)

This is also known as desalination or chloride removal. Many applications on chloride-contaminated structures in the UK have been experimental. The life and effectiveness of the treatment is still being assessed, and it is likely to be structure-specific in terms of concrete quality, cover, reinforcement details, chloride content and exposure. In principle, ECE does not need permanent power supplies or monitoring, although in practice checks on the corrosion condition are recommended to determine the life of the treatment.

The technique uses a similar set up to the ICCP system shown in Fig. 5.6. The difference is that ECE uses a temporary anode and passes a high current (about $1\,A/m^2$ of steel or concrete surface area) to pull the chlorides away from the steel. A proportion (usually around 50–90%) of the chloride can be removed from the concrete, with a very significant removal immediately around the steel and a high level of repassivation of the steel.

ECE can be used in many of the situations where cathodic protection can be applied. It is at its best where the steel is reasonably closely spaced, where the chlorides have not penetrated too far beyond the first layer of reinforcing, and where future chlorides can be excluded. It has been applied to highway structures, car parks and other structures in Europe and North America.

The treatment takes about 8 weeks. On completion, the anode is removed and there are no on-going monitoring requirements. If sufficient chloride is removed and further chlorides are excluded, then it is a one-off treatment.

Some research has shown an improvement in concrete properties such as freeze-thaw, chloride, carbonation and water uptake resistance (Buenfeld & Broomfield, 1994). However, this research has not been replicated in other laboratory tests or with other concrete mix designs.

The ECE system has various advantages and limitations compared with cathodic protection and other treatments. The treatment time can be too long to make it feasible. The problems with isolated steel, prestressing and ASR mentioned for ICCP are exacerbated for ECE owing to the higher voltages and current densities. Some concerns with reduction in bond have been identified, but these only apply to structures containing smooth reinforcement where the mechanical interlock of ribbed steel is not available. A lithium-based electrolyte has been tested to mitigate the ASR risk, but the results on its effectiveness are not yet conclusive.

Realkalisation

This can be described as the equivalent of ECE for structures where the carbonation has reached unacceptable depths. The treatment is quicker to carry out than ECE (days instead of weeks), and the proprietary system uses a carbonate electrolyte as an aid to regenerating the alkalinity in the concrete and around the rebar. The same anode systems are used for real-kalisation as for ECE.

The treatment is most commonly applied to carbonated buildings with carbonation damage but with at least moderate cover to the reinforcing steel (greater than about 10 mm). The Norcure™ list of proprietary real-kalisation treatments carried out under license shows a total of 182 452 m² treated from 1987 to 1995. The merits and limitations of realkalisation are basically similar to those of ECE, but the shorter treatment time and lower charge passed mitigate some of the concerns.

Structural strengthening

Structural strengthening may be required when the structural appraisal has revealed that the structure requires up-grading in order to perform its existing function or for a proposed new use. There are a number of means of achieving increased structural performance, including (DD ENV 1504-9, 1997):

- additional reinforcement bars (external or embedded);
- adding bonded reinforcement bars;
- composite plate bonding;
- the addition of concrete or mortar;
- injecting cracks, voids or interstices;
- prestressing.

Structural strengthening requires design, specification and overseeing by a suitably qualified engineer.

Crack repair

As discussed in Chapter 4, there are a number of reasons why cracking occurs, and it is imperative that the cause of the cracking is correctly diagnosed so that the appropriate repair strategy can be identified. It is the crack width and the quality of the concrete cover that determines the necessity for repair. It is also important to determine whether the crack is dormant or active.

There are a number of different systems for repairing cracks, which can be categorised as (Standards Australia, 1996):

- filling;
- sealing (to prevent ingress of water and other agents of decay);
- stitching;
- elastic joints (creating a joint at the crack with some elasticity);
- load-bearing connection (creating a tension-proof joint to restore load-bearing capacity).

Hairline cracks and small cracks that are dormant do not require repair for technical reasons and may self-seal. Cracks that are dormant and clean can be filled with a material that is compatible with the concrete, whilst active cracks must either be repaired by arresting the movement and filling as for dormant cracks, or by using a flexible filler that will accommodate the on-going movement. Once cracking has allowed the passage of agents to facilitate corrosion of the reinforcement, the repair will need to include repairing the corroded reinforcement.

Crack filling

There are a number of requirements for any material used to fill cracks, including:

- low viscosity to penetrate the crack;
- good workability;
- mix stability;
- ageing resistance;
- strength;
- compatibility with substrate;
- adequate adhesion;
- low volatile content.

The selection of the material will also be dependent on whether the crack is dormant or active, the width of the crack and the practicalities of the repair. The injection method is used to fill cracks down to 0.05 mm wide. Holes are drilled at intervals along the crack and epoxy or polyurethane is injected under high pressure. This is a specialist repair technique and experienced contractors are needed. Cementitious products can be used to fill wider cracks.

Crack sealing

Dormant cracks, where the repair does not have to perform a structural role, can be repaired by enlarging the crack along the external face and filling and sealing it with a suitable joint sealer. This method is commonly used to prevent water penetration to cracked areas. Various materials are used, including epoxies, urethanes, silicones, polysulphides, asphaltic materials and polymer mortars. The crack is routed out, cleaned and flushed out before the sealant is placed (ACI 224.1R-93, 1993).

Stitching

Stitching can be used where it is necessary to re-establish tensile strength across a cracked section of concrete. Stitching is achieved by stapling across the crack in a manner that spreads the tension across a larger area. However, this method can stiffen the structure and cause cracking elsewhere, and therefore additional strengthening of adjacent sections may be required.

Other methods

Grouting, dry-packing, impregnation, overlays and surface treatments, and autogenous healing are all methods of crack repair that can be appropriate in certain situations. Autogenous healing is a naturally occurring process that may occur in dormant cracks that are wet but are not subject to water flow. It works as a result of the continued hydration of the cement and the carbonation of the concrete. The precipitation, accumulation and growth of the calcium hydroxide and calcium carbonate crystals chemically bond and restore some integrity across the crack (ACI 224.1R-93, 1993).

A common misconception is that surface coatings have the ability to bridge cracks and therefore protect the building from ingress by aggressive agents and water. Coatings should not be relied on to bridge cracks other than dormant hairline cracks. Expert advice should be taken when considering appropriate crack repair strategies. For active cracks most repair methods will have a limited life, and the cause of the problem will need to be identified and rectified if possible to prevent on-going concrete deterioration.

Corrosion monitoring

Monitoring may be installed to check for changes in the condition after any of the above treatments, including doing nothing, and can be used around the edges of repairs or concrete replacement. Corrosion monitoring has been applied after treatments with sealers, inhibitors and ECE (see Chapter 4).

Life-cycle cost analysis and technical comparison of treatments

For any structure with more than a few years residual service life, it is normal practice to carry out life-cycle cost analysis to determine the most cost-effective repair and future maintenance options. This is structure-specific and often client-specific, depending on the budgets, accounting procedures and commercial attitude to the structures owned.

Long-life structures (20–100 years of residual service life) usually justify the higher initial cost of 'global' treatments such as ICCP. For shorter-life-span structures, issues such as access or the acceptability of regular repair cycles often dominate the choice of treatment. Life-cycle costing is discussed more detail in Chapter 6.

Trials

It is likely that trials will be required prior to initiating the repair works, whatever system of repair is selected. Trials can play an important role at the specification development stage of the project, or may be carried out early on in the contract and used to agree on the repair methods selected, the standard of work, and any other details such as the level of cleaning, colour and texture of patch repairs, and so on.

In historic building work, it is not uncommon to commission a small trial contract during the specification development stage to assist in determining the appropriate method of repair and establish the most suitable materials for the job. This will obviously be influenced by whether access is available and the client's needs in terms of the precision of the tender. Any practicalities can also be sorted out at this time. If any new repair methods are to be attempted or developed, then it is preferable to have established what these are prior to the letting of the main contract. This ensures that the specification is more authoritative, and means tenders will be based on known costs. However, trials may not completely eliminate the need for variation during a contract, and some flexibility will need to be built in.

Where trials cannot be carried out at this stage they will need to be built into the tender. If the precise repair method has not been determined, then the specification can include a range of potential repair methods, and schedules of rates can be required as part of the tender for the various alternatives to ensure that the tenders can easily be compared. If the methods cannot be fully specified at this stage, method statements can be requested as part of the tender, which will allow the tenders to be compared properly, and may also encourage the contractor to develop a more creative approach to the specific issues.

Repairing listed buildings and structures

As discussed in Chapter 1, listed buildings and structures may have additional statutory controls that influence the repair strategy. One of the key principles used to develop a conservation strategy is that of minimum intervention: this advocates that only repair work that is absolutely necessary should be carried out, and the work should not affect the architectural values of the place. However, concrete repair is an invasive process, particularly once latent damage has occurred, and requires larger-scale intervention than is usually necessary for traditional masonry structures. In the case of a brick or stone building, for instance, individual components can be replaced or repaired relatively easily without affecting the structural integrity of the building. *In situ* concrete, being monolithic in form and structure, cannot be dealt with in the same way. There are also well-established repair methods that have been developed for historic masonry structures, whereas conservation techniques for concrete repairs are still largely undeveloped. The impacts of the various repair methods on exposed concrete buildings are summarised in Table 5.2, and this highlights some of the difficulties associated with conserving buildings of this type.

As mentioned previously, once visible damage has occurred it is difficult to match patch repairs to the parent concrete. The usual practice of utilising proprietary mixed repair products with their higher technical performance may not provide a suitable visual match for existing concrete. Even if a good match is achieved, the patches may weather differently to the parent concrete. This is particularly problematic when the surface finish of the concrete is an important contributor to the significance of the building. Coatings are often recommended as part of the repair strategy, and whilst they hide patch repairs, they can drastically alter the appearance of the structure. However, there have been a few instances where listed buildings controls have required repairs to be carried out in a manner that minimises the visual impact on the building (Fig. 5.8). In the United States, as the Promontory Apartment case study in Chapter 8 shows, it is generally accepted that specially batched concrete repair materials will be required to achieve a repair that does not have a visual impact on the building.

Table 5.2 Typical reinforced concrete decay problems for exposed concrete buildings, current repair options and conservation dilemmas.

Cause of decay	Manifestation	Repair solution*	Conservation dilemma
1.0 Inherent material problem			
1.1 Low cement content and finely ground cement	Highly permeable concrete: poor durability leading to reinforcement corrosion	Traditional repair and coating Realkalisation and coating Cathodic protection Corrosion inhibitor Sprayed cementitious over coating/render	Loss of original material, change in appearance (unless patches match original concrete), coating changes surface appearance Some physical damage from application of process, coating changes appearance Potentially none Potentially none New surface to building
1.2 High-alumina cement	Gradual loss of strength leading to structural failure	No cure	
1.3 Poor-quality aggregates: Impure aggregates (unwashed sea sands) Poorly graded/shaped aggregates	Poor durability Chloride attack Poor workability, high water/cement ratios leading to poor durability and reinforcement corrosion	Desalination As 1.1	Some physical damage from process As 1.1
1.4 Alkali-aggregate reaction	Concrete breakdown	No cure	Potentially none
1.5 Presence of calcium chloride additives in mix	Chloride attack	Cathodic protection or Desalination	Some physical damage from process
1.6 Creep	Structural failure Surface cracking	No cure Crack filling	Aesthetically intrusive unless fillings match original concrete Potentially none Some physical damage from process
1.7 Decorative finishes: Acid etching Bush hammering/surface profiling	Reduced alkalinity of concrete, increasing susceptibility to reinforcement corrosion Reduced cover to reinforcement	Cathodic protection or Realkalisation As 1.1	As 1.1
1.8 Poor quality/lack of reinforcement, scrap and inconsistent type	Different levels of cover Structural failure	As 1.1 No cure	As 1.1

2.0	**Environmental influences**		
2.1	Acid gases	Poor durability leading to reinforcement corrosion	As 1.1
2.2	Air and moisture	Poor durability leading to reinforcement corrosion	As 1.1
2.3	Freeze–thaw	Poor durability leading to breakdown of concrete	Coating / Sprayed cementitious over coating/render	Coating changes appearance / New surface to building
2.4	Seawater	Chloride attack	Cathodic protection and/or Coating	Potentially none / Change in appearance
2.5	Road salts	Chloride attack	Cathodic protection and/or Coating	Potentially none / Change in appearance
2.6	Sulfate attack	Concrete breakdown	Coating	Change in appearance
3.0	**Poor design and workmanship**		
3.1	Lack of cover to reinforcement/placement	Poor durability leading to reinforcement corrosion	As 1.1	As 1.1
3.2	Remote site batching—poor mix quality	Poor durability leading to reinforcement corrosion	As 1.1	As 1.1
3.3	Inadequate control of water content	Poor durability leading to reinforcement corrosion	As 1.1	As 1.1
3.4	Inadequate curing	Cracking, loss of strength, and poor durability	Crack filling and elastomeric coating	Change in appearance
3.5	Inadequate compaction/vibration	Cracking, loss of strength, and poor durability	As 3.4	As 3.4
3.6	Plastic shrinkage	Cracking, loss of strength, and poor durability	As 3.4	As 3.4
3.7	Accessibility for maintenance	Lack of maintenance	Dependent on how manifested	
3.8	Design faults	Water ponding, choice of fixings of attached materials, weathering details, etc.	Improvements to detailing	May involve alteration to building's appearance

* The repair will often be a combination of one or more of the methods listed.

Fig. 5.8 The concrete repairs at the Alexandra Road Housing Estate in Camden were carried out in a manner that attempted to retain the original concrete finish and appearance. Previous repairs had resulted in numerous patches all over the building that were not a good match with the parent concrete (photo Susan Macdonald).

As listed building consent will be required for repair works to listed buildings, it is important to provide evidence of a thorough analysis of the cause of any problems, and to show how the repair strategy has been developed to accommodate all of the usual requirements, including conservation.

An analysis of the significance of the building is an important first step in the conservation planning process. This will assist in identifying appropriate repair methods, assessing the impact of the proposed repairs on the structure's significance, and determining the longer-term maintenance approach. It will also help balance the long-term affects of the proposed repairs against any negative impacts on the significance of the building. It may be, for example, that a one-off repair solution will not be the most suitable approach, and a longer-term managed programme of repair and maintenance may prove the best option.

A useful exercise to carry out when approaching a conservation project is to weigh up all the options, from do nothing to major intervention, against life-cycle costs and repair impacts to devise an appropriate approach.

Organisations such as English Heritage, Docomomo and APT have held a number of conferences and seminars where the particular conservation issues raised by concrete repair have been discussed. These, and a number of other useful texts on this subject, are listed in the further reading section at the end of this chapter.

Selecting a contractor

The success of any repair or remediation project lies with the consultants that assess the building and develop the appropriate repair strategies, as well as the contractor who implements the specification on-site. A good working relationship between the specifiers, be they an architect, engineer, surveyor or specialist concrete consultant (or any combination of these), and the contractor is essential (Ashurst, 1994). The team developed for a particular project should have an agreed common aim and work together to achieve this. Enlisting the help of a suitably experienced contractor during the specification development stage of a project can be a very useful way of ensuring that the specification is practically applicable and will achieve the anticipated outcome.

As discussed in Chapter 3, the consultants responsible for assessment, appraisal and specification development should be appropriately qualified and experienced for the task. Likewise, the contractor should be suitably experienced in the particular methods of concrete repair to be employed on the project. It is useful to visit previous projects carried out by potential contractors and ask for references from clients. Interviews with potential contractors can be an important part of the tender process and can test the suitability of the contractor for the project. Tender lists should include contractors of comparable skills and experience so that meaningful comparisons can be made between the tenders received (Ashurst, 1994). Contractors unskilled in carrying out work included within the specification are unlikely to provide a realistic price for the work.

It is important to know who will be supervising the contractors on-site, and be satisfied they are appropriately experienced and skilled for the task. The proposed workforce should be employees of the contractor, or at least be employed for the duration of the project, and all should be suitably experienced for the proposed works. If some of the work is to be sub-contracted, it is important to ensure that the sub-contractors are known as part of the tender, and that the consultant administering the tender process is satisfied that the sub-contractors proposed are appropriate for the job. The contractor should be able to demonstrate their commitment to the relevant Health and Safety Act requirements, and have suitable on-site policies and practices.

There are various trade and professional bodies associated with concrete repair that have accredited lists of members for the various methods of concrete repair; these are listed in Appendix A.

There are very few concrete repair contractors who have experience in dealing with the particular requirements associated with listed buildings work. In this case, the tenderer's attention should be drawn to any particular requirements, and the specifier needs to stress the importance of adherence to these requirements when inviting or interviewing tenders. This said, there are also few conservation practitioners who have experience in repairing concrete buildings, and so teamwork will be vital to the success of the project.

As mentioned in Chapter 3, concrete construction has historically been in the hands of specialist contractors, and this has also been the case with concrete repair works. Many repair systems—such as realkalisation and chloride extraction—are licensed to particular contractors. There has been a tendency for owners to approach repair companies, or even product manufacturers, directly for the assessment and repair of their buildings rather than engage the help of an engineer or architect for repairs. This can mean that the contractor or manufacturer will invariably specify a repair that is limited by their own range of products, or the methods of repair they are able to perform, and the cause of the problem may not have been correctly diagnosed. This approach is unlikely to be beneficial for the building; the repairs may be ineffective in dealing with the problem, and result in larger-scale and more expensive repairs at a later stage.

Recording the works

It is essential to keep good records of any repair or maintenance work, both as specified and as it was carried out on-site. Once work commences on-site, specifications can change as unforeseen issues arise, and it is important to ensure that records are kept documenting any changes to the specification and recording why these occurred. The information gathered as part of the assessment exercise, as discussed in Chapter 3, should also be retained along with these records. These records will inform future maintenance and repair works as well as being useful in monitoring the repairs carried out, and therefore they must be made available to the owner/manager rather than remaining with the consultants or the contractor. Record keeping should be an integral part of the consultant's and contractor's work, and allowances should be made for this in their contracts.

References

ACI 224.1R–93 (1993) *Cause, Evaluation and Repair of Cracks in Concrete Structures.* American Concrete Institute, Detroit, MI.

Ashurst, N. (1994) *Cleaning Historic Buildings. Vol. 2. Cleaning Materials and Processes.* Donhead, Shaftesbury.

Buenfeld, N.R. and Broomfield, J.P. (1994) *Effect of chloride removal on rebar bond strength and concrete properties.* Corrosion and corrosion protection of steel in concrete, p. 1438–50. Sheffield Academic Press.

Concrete Society Technical Report No 26 (1985) *Repair of Concrete Damaged by Reinforcement Corrosion.* Concrete Society, Slough.

DD ENV 1504-9 (1997) *Products and Systems for the Protection and Repair of Concrete Structures. Definitions, Requirements, Quality Control and Evaluation of Conformity. Part 9. General Principles for the Use of Products and Systems.* British Standards Institution, London.

Standards Australia (1996) *Guide to Concrete Repair and Protection.* Standards Australia, Standards New Zealand, ACRA, CSIRO, Sydney.

Further reading

ACI 222R-96 (1996) *Corrosion of Metals in Concrete.* American Concrete Institute, Detroit, MI.

Broomfield, J.P. (1997) *Corrosion of Steel in Concrete: Understanding, Investigation and Repair.* E & FN Spon, London.

BSEN 12696 (2000) *Cathodic Protection of Steel in Concrete.* British Standards Institute, London.

BSEN prTS14038-1 (2002) *Electrochemical re-alkalisation and chloride extraction treatments for reinforced concrete. Part 1: Re-alkalisation.* British Standards Institute, London.

Chess, P.M. (ed) (1998) *Cathodic Protection of Steel in Concrete.* E & FN Spon, London.

Concrete Society Technical Report No. 36 (1989) *Cathodic Protection of Reinforced Concrete.* Concrete Society, London.

Coney, W. (1996) *Preservation Brief No. 15. Preservation of Historic Concrete: Problems and General Approaches.* US Department of the Interior, National Park Service, Washington, DC.

CPA (1995) *Cathodic Protection of Reinforced Concrete. Status Report No. SCPRC/001.95.* Corrosion Prevention Association, Aldershot.

Currie, R. and Robery, P. (1994) *Repair and Maintenance of Reinforced Concrete.* British Research Establishment, Watford.

De Jonge, W. and Doolaar, A. (1997) *The Fair Face of Concrete: Conservation and Repair of Exposed Concrete.* Preservation Dossier No. 2, DOCOMOMO International, Delft, The Netherlands.

EN 12696: 2000 (2000) *Cathodic Protection of Steel in Concrete.* European Standard, British Standards Institution, London.

European Federation of Corrosion Publication 24 (1998) *Electrochemical rehabilitation methods for reinforced concrete.* (Ed. J. Mietz) Published for EFC by IOM Communications, London. ISSN 1354-5116.

Gibbs, P. (2000) *TAN 20. Corrosion in Masonry-Clad Early 20th-Century Steel-Framed Buildings.* Historic Scotland, Edinburgh.

Macdonald, S. (ed) (1996) *Modern Matters: Principles and Practice in Conserving Recent Architecture.* Donhead, Shaftesbury.

Macdonald, S. (1997) Authenticity is more than skin deep: conserving Britain's post-war concrete architecture, and other articles on concrete repair in *Mending the Modern.* Special issue of the APT Bulletin, Vol. XXVIII, No. 4.

Macdonald, S. (ed) (2001) *Preserving Post-War Heritage: The Care and Conservation of Mid-Twentieth Century Architecture.* Donhead, Shaftesbury.

NACE Standard RP0290 (2000) *Impressed-Current Cathodic Protection of Reinforcing Steel in Atmospherically Exposed Concrete Structures.* National Association of Corrosion Engineers (NACE International), Houston, TX.

NACE International Publication 01101 Item No. 24214 (2001) *Electrochemical Chloride Extraction and Re-alkalisation of Reinforced Concrete.* Part 1: Electrochemical Chloride Extraction from Steel Reinforced Concrete – A-State-of-the art report. NACE International, Houston, USA.

Part Three
Maintaining a Concrete Building

Chapter 6

Maintenance of Concrete Buildings and Structures

Stuart Matthews, Matthew Murray, John Boxall, Ranjit Bassi & John Morlidge

Introduction

Concrete has historically been seen as material that will not deteriorate. This has engendered the once widely held misconception that there was no need to effect repairs or undertake any maintenance. Whilst many concrete structures have performed well and have given good long-term performance, experience has shown that this is not so for them all. There have been many examples of premature deterioration of reinforced concrete structures, particularly in aggressive chloride-laden environments.

Maintenance measures for concrete structures and buildings, i.e. where periodic or on-going actions are required to maintain their current functionality, may include:

- cleaning of surfaces;
- application of coatings, sealers or surface impregnation to provide protection against an aggressive environment or deterioration of the concrete surface;
- periodic replacement of sealers and fillers in movement joints;
- the operation of cathodic protection systems;
- the application of corrosion inhibitors;
- replacement of sacrificial layers in circumstances where abrasion or erosion occurs.

This chapter provides practical guidance on issues relating to the maintenance of concrete structures, particularly in respect of the cleaning and coating of concrete surfaces. The main focus is upon reinforced concrete structures and buildings. However, it is important to view maintenance activities in a wider context, and this is done by considering whole-life costing and life-cycle performance issues.

Definitions

The terms design-service life, economic-service life, whole-life cost and so on have specific meanings, and definitions are provided in the Glossary where indicated. It should be recognised that these concepts are still developing as the topics are debated and guidance is prepared. Other authors and publications may cite other criteria or definitions. The end of service life in accordance with a particular criterion is deemed to be when the structure is unable to meet the specified performance requirements, or when a previously defined unacceptable state is reached. Work remains to be done to develop the life-cycle approach into a mature philosophy with associated methodologies for its effective and commonplace implementation.

Concepts of whole-life performance

Many building owners, managers and construction professionals, after experience of the repair of concrete structures during the last decade or so, are now more aware of the need to address whole-life and residual-service life cost (see Glossary) and associated engineering issues in a more holistic and structured way. This type of approach should enable them to understand more clearly all the relevant costs and revenues associated with the acquisition and ownership of an asset.

Care should be taken to avoid falling into the trap of considering these matters simply as an accountancy exercise. The financial aspects are perhaps best thought of as tools that allow comparisons of alternative construction and management options. Of greater importance are the philosophical issues relating to the life-cycle of an asset, and the importance that is associated with service-life design concepts. There are clearly links with the current desires to achieve more sustainable forms of construction. Best conservation practice has long advocated this approach. Accordingly, it must be expected that with time there will be a shift away from considering the costs of construction only.

It is also important to realise that decisions made or actions undertaken at different stages in the life-cycle of a structure or building have vastly different impacts upon its whole-life cost and performance. As the life-cycle proceeds through the concept, planning, design and specification phases, to construction and use, and finally to disposal, the relative influence of decisions and actions taken at particular stages progressively diminishes. It has been stated that as much as 95% of the whole-life cost of a building may effectively be set by the time the building is complete (HB 10141, 1997).

The often-quoted 'de Sitter's law of five's' illustrates the effect of decisions made at different stages in the life-cycle of a building or structure: paraphrasing, this might be expressed broadly as follows for a reinforced or prestressed concrete structure:

> £1 spent getting the structure designed and built correctly is as effective as £5 spent in subsequent preventative maintenance in the pre-corrosion phase while carbonation and chlorides are penetrating inwards towards the steel reinforcement. In addition, this £1 is as effective as £25 spent in repair and maintenance when local active corrosion is taking place, and this is as effective as £125 spent when generalised corrosion is taking place and where major repairs are necessary, possibly including replacement of complete members.

This emphasises the importance of the concept, planning, design and specification phases, and the major influence they have upon whole-life cost and life-cycle performance. Accordingly, it is very important to make the right decisions early in the life of a building or structure.

Increasingly, owners and managers are also becoming concerned with the certainty of achieving agreed cost and specified performance parameters. In this they are seeking to control the risk to which they are exposed by a decision to commission new construction or refurbishment works. Of course the first cost—the construction and commissioning—must be right. More importantly, owners wish to avoid premature failure; not only structural collapse, but any circumstances that restrict their ability to use the facility and so disrupt their business processes. If maintenance works will be required, such as the renewal of surface coatings every 10 years or so, they want to know this so that they can plan and budget for it. In this way, performance considerations will increasingly influence the concept of service life by design, which includes maintenance requirements.

Achieving a satisfactory life-cycle performance for a proposed reinforced-concrete structure will depend on a number of technical factors, including those cited below. In recent years, the principal problems with reinforced- and prestressed-concrete structures have concerned their lack of durability. Accordingly, consideration should be given in the planning, design and specification stages to issues such as those listed below.

- Correct assessment of the loading regimes and environmental conditions to which the concrete will be exposed, particularly where this involves exposure of reinforced or prestressed concrete to chlorides. These factors include the general environment conditions (macroclimate), and influences such as the location and orientation of the concrete surface being considered and its exposure to prevailing winds and rainfall (meso-climate), as well as localised conditions such as surface ponding, and exposure to surface run-off and spray (micro-climate).

- Making a rational decision on the required design and service lives for the structure. This includes consideration of the serviceability needs of the structure, i.e. can some deterioration be tolerated, or must the concrete remain pristine within the intended service life? Is uninterrupted use of the facility of paramount operational importance?
- Correct design and specification of the concrete and the structure including, in particular, its 'buildabiilty'.
- Good construction practice.
- Appropriate in-service inspection, monitoring and maintenance actions.
- Correct and early assessment of any deterioration, followed by effective repair or preventative measures.

Design and service-life states

A number of alternative approaches to service-life design can be employed. These include the two described below.

- Prescribed (deemed to satisfy) approaches of current codes for structural design, such as those set out in BS8110 (1985) and DD ENV 1992-1, Eurocode 2 (1992). Although simple in format and widely used, this approach may not be effective and is sometimes unhelpful. Technically, the approach may be questionable in some circumstances as it does not help to achieve a specific service life, except in a generalised way.
- Performance-based approaches being advocated in recent standards and technical publications; current concepts for defining design and service-life states include factor-based methods (HB 10141, 1997) and probabilistic approaches (Siemes and Rostam, 1997; CEB Bulletin 238, 1997).

The move towards the concept of performance-based serviceability design requires, amongst other things, an ability to decide when a particular serviceability or safety criterion has been reached. As part of the approach, it is necessary to define appropriate criteria that could be used as a basis of this judgement. These can be developed from the various states or conditions that might arise during the life of a structure or building. Recognition needs to be given to the degradation processes that could affect the ability of an individual element, or the overall structure, to

perform one or more of their intended functions. These states are typically defined in terms of key performance or functional requirements. A
number of definitions are presented in the Glossary for a range of possible service-life circumstances. These definitions give some understanding of the potential complexity of the approach, and the sophistication it
might allow in assessment and decision making.

For example, the end of the technical service life (see Glossary) of a
reinforced-concrete element could be taken as one of a number of
possible criteria. Depending on circumstances, these might include:

- the point when corrosion is initiated;
- the first appearance of cracking (visible with magnification);
- cracking visible to the naked eye;
- first spalling;
- the development of excessive deflection under a particular loading
  configuration;
- when the load capacity assessed is lower than a predetermined value;
- when the probability of failure under the design or other specified
  loading exceeds a predetermined value, perhaps expressed as a reliability index.

It can be seen that in this instance the technical service lives defined relate
to increasing degrees of deterioration of the concrete element.

This approach to service-life design does not result in a single value,
but recognises that it may vary depending on a number of factors, including the type of element or structure and the associated performance
requirements, as well as on the maintenance regime that is to be adopted.
In addition, environmental and aesthetic aspects can strongly influence
considerations about what comprises acceptable technical performance,
and these are parameters that may need to be addressed.

The concept of performance-based serviceability design also requires:

- a means of evaluating the ability of a structure or building to meet the
  specified (minimum) standards of acceptable technical performance;
- methods for evaluating the changes that occur over time in the ability
  of the structure to meet the specified/acceptable technical performance standards.

In most instances, the tools required to do this are still being developed
and refined. Accordingly, it is expected to be some time before these
approaches are widely used and employed to define when maintenance
or remedial activities should be performed.

The influence of surface features upon the weathering and appearance of concrete structures

Despite the introduction of concrete technology in the nineteenth century, its use as we know it today took some time to implement. The new architectural forms of the 1920s and 1930s, particularly those used by architects and designers interested in the Modern Movement, embraced the potential of concrete to provide new undecorated forms. The abandonment of traditional weathering details to create a new streamlined aesthetic, together with the lack of knowledge of the best practice to ensure long-term durability, had serious implications for concrete structures from this period. Lack of detailing to ensure that water is shed from these buildings has resulted in patchy uneven soiling of many concrete buildings. Textured finishes, such as those introduced in the post-war period, have exacerbated these problems.

Today there is renewed interest by architects and property managers, amongst others, in the cleaning of concrete to maintain its appearance. This is coupled with a growing application of surface coatings to enhance the appearance of buildings and structures. It is expected that consideration of the long-term maintenance or conservation of historic concrete buildings and structures will become increasingly important as more receive recognition and legal protection through listing and similar measures.

Options, approaches and techniques for maintaining concrete buildings

Most reinforced-concrete buildings exposed to normal environmental conditions and imposed loading regimes are unlikely to suffer from significant deterioration within their design-service life. However, there are numerous examples where a combination of factors, including poor initial design and detailing, inadequate consideration of specific environmental loadings, lack of supervision and bad workmanship during construction, have led to conditions permitting corrosion of the reinforcement. As explained in Chapter 4, the expansive forces arising from the corrosion products can result in cracking and staining of the concrete, or in spalling of the cover concrete and a possible reduction in structural capacity. This is perhaps the most common form of significant premature deterioration of concrete structures.

The major causes of reinforcement corrosion are carbonation of the cover concrete and the ingress of chlorides in the presence of moisture. There is a greater risk of corrosion in reinforced-concrete structures exposed to more severe environmental loading, especially those subject to marine and de-icing salt conditions.

These and other potential mechanisms of deterioration have been reviewed in Chapter 4. Once signs of deterioration are evident, it is generally desirable that remedial action on the structure or building is taken promptly. Delay may considerably increase the cost of subsequent maintenance or repair works.

The recently published *British Standard Draft for Development* (DD ENV 1504, 1997, Part 9) groups prospective repair and remediation methods on the basis of the principle being utilised. ENV 1504 Part 9 is set to become the obligatory method for specifying concrete repair works. An overview of the approach and options embodied within Part 9 of the standard is depicted as a flow chart in Fig. 6.1. This identifies information requirements, and options and factors influencing different stages of the repair and remediation process. Any method that can be shown to achieve compliance with the chosen principle of repair and protection may be considered. Figure 6.1 helps to illustrate the relationship between the repair works as discussed in Chapter 5 and the requirements for future maintenance activities.

The principle(s) and method(s) chosen for repair and protection should:

- be appropriate to the type and cause, or combination of causes, of deterioration and to the extent of the defects;
- be appropriate to the proposed length of the residual-service life, and the anticipated environmental conditions and structural loading in this period;
- be compliant with the process chosen to control the deterioration mechanism(s) identified;
- recognise the availability of products and systems that are suitable for the repair of the structure (in due course these are likely to need to comply with the requirements of the DD ENV 1504 series or other relevant EN or European Technical Approval documents);
- be compatible with the protection or repair option(s) chosen for other components in the structure.

There are other circumstances where concrete may require particular repair or maintenance actions. Whilst it is impractical to provide a definitive list of all these circumstances or the types of structure that may be involved, potential examples could include:

- floor and ground slabs where use might create problems of surface dusting, abrasion/erosion, edge damage under heavy industrial use, surface damage under freeze–thaw actions, etc.;

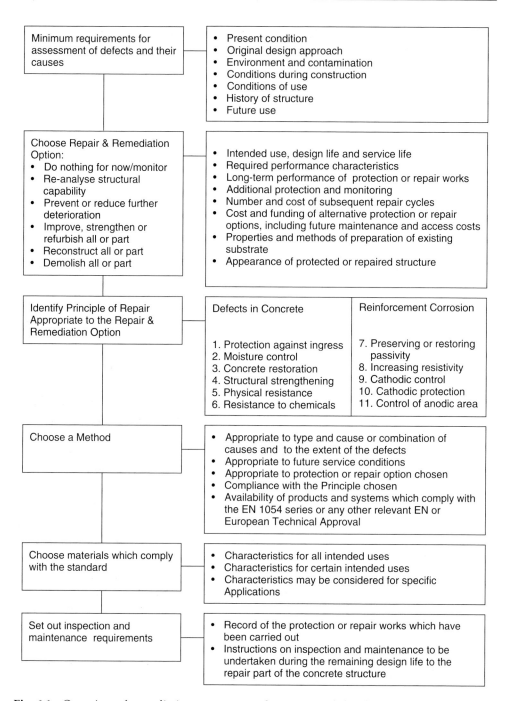

Fig. 6.1   Overview of remediation processes and options as defined in DD ENV 1504 (1997, Part 9).

- concrete subject to chemical attack, or requiring enhanced resistance in a chemical plant or similar circumstances;
- maritime structures;
- bridges;
- car-parking structures.

The following list illustrates some of the maintenance works that might be undertaken on a building or structure:

- cleaning of surfaces;
- application of a surface coating;
- application of surface impregnation;
- the periodic replacement of sealers and fillers in movement joints;
- application of corrosion inhibitors;
- installation/renewal of sacrificial layers in circumstances where abrasion/erosion of surfaces occur;
- operation of a cathodic protection/prevention system;
- ensuring the effective operation of surface water drainage systems, movement joints and the like.

There appear to be a range of views and current practices as to what constitutes a maintenance activity and what is considered as a repair action. Those techniques that can be categorised as essential repairs have already been discussed in Chapter 5. Clearly, items such as coatings and surface impregnants, typically introduced to provide protection or enhanced resistance against the effects of exposure to an environment aggressive to either concrete or reinforcement, have a finite service life and will need periodic renewal.

The importance of on-going general maintenance actions that can address the cause of the problem, such as controlling the discharge of surface water containing chlorides to avoid unnecessary contamination of parts of the concrete structure, should not be underestimated. For example, leaking movement joints in some bridges have introduced chloride contamination into half-joints and bearing zones. In a number of instances, chlorides have built up over time to appreciable levels, promoting severe corrosion of reinforcement in these particularly sensitive locations.

The management strategy for a building or structure may need to take account of factors such as:

- intended use of structure;
- required residual-service life of the structure;
- required performance characteristics, including any change of use or appearance.

The approach adopted should include consideration of the need for long-term maintenance, the requirements for proper inspection of key areas of the structure's components, and the installation of appropriate monitoring procedures. Early warning of impending problems would allow proper management and timely development of remedial measures that should ensure better control and minimisation of overall expenditure.

For a proper maintenance programme to be implemented, it is essential that detailed records are kept of the repair or protection works that have been carried out. Similar records should be available for inspections and technical assessments. This is sometimes prepared in the form of a technical log for a structure or building. Instructions on inspection and maintenance to be undertaken during the residual-service life (see Glossary), both of the repaired part and the remainder of the concrete structure, should be laid out in a form that allows it to be understood and implemented easily. The maintenance plan/schedule needs to define what should be done, where, when, by whom and possibly also how.

As discussed in Chapter 5, it may be inappropriate to apply a single repair method to the whole structure. Often, different elements of the structure or building will have experienced variable environmental exposure conditions and microclimates, or may have suffered from one or more of the contributory deterioration mechanisms. Each element or group of elements could therefore require an individual approach to its repair and protection. A similar approach could be required for the treatment of maintenance issues.

Cleaning soiled concrete surfaces

Cleaning concrete typically covers two scales of economy. At one end there is fast and cheap cleaning of less prestigious concrete surfaces for immediate economic purposes. At the other end there is the consideration of long-term maintenance or conservation of new or historic concrete materials and buildings.

There is currently a strong new enthusiasm for cleaning less prestigious buildings with the aim of urban regeneration. A fresh new appearance may contribute confidence, which might stimulate inward investment, or be the basis of a zero-tolerance policy. Cleaning a building is essentially a cheap and effective component of local economic regeneration. There can, therefore, be a wide range of reasons to clean all types of concrete buildings.

If the building is historic or prestigious, then a wider appraisal of the cleaning operation requirements and associated constraints may be required. Listed buildings may require statutory approval for cleaning, and the local authority or relevant national heritage body (such as English Heritage, Historic Scotland, etc.) should be contacted prior to proceeding.

Identification of the concrete surface

In order to clean concrete effectively, the nature and condition of the surface material must be identified. Concrete is a mixture of cement and aggregates of various sizes. The face of concrete that has set naturally in air or against a mould will develop a 'skin'. At the surface there are smaller aggregate particles, and a distribution of the finer ones. If the concrete is cut, then the surface revealed will always have different physical properties and appearance from the natural set surface.

If the aggregates are extensively visible as hard, pebble-like stones, then either there is some existing surface decay (i.e. the loss of the surface laitance), or the surface has been designed with an exposed aggregate finish. These different surfaces may require different methods of cleaning. Pollution soiling adhering to the aggregate particles may be different to the soiling attached to exposed cement matrix. Some aggregates contain carbonate material which restricts the use of acidic cleaners.

When concrete is made with a substantial amount of stone particles as the aggregate, it may effectively be an artificial stone. Some modern decorative concrete cladding will not even necessarily look like conventional concrete. Delicate pebble-dashed appearances can be produced by specialist production methods, and pigments can also be added as a fine, coloured aggregate. Where a decorative concrete finish is present, there is often a requirement to maintain it. Examples are the patterns of wood mouldings and smooth finishes that form part of the design of the building. The skin can be weathered or more porous, and so may be susceptible to damage from cleaning methods considered appropriate for exposed aggregate. The cement matrix may require a separate, much more delicate and more expensive, approach to cleaning.

Soiling identification

To select a single cleaning method, or range of methods, the type of soiling on the building must be identified. This will require close examination of all surfaces. For example, black pollution staining may be confused with dark lichen growth, and these will require very different methods of cleaning.

Trial cleaning

Ideally, trials of various methods of cleaning are carried out to determine the best method. The trial cleaning may be included as part of a specification for cleaning for each particular soiling type and surface finish of the structure. In most cases, the advice of a specialist or an experienced contractor should be sought to determine the most appropriate method in the circumstances. Trials were discussed in more detail in the previous chapter.

Cleaning preparation

In some cases it may be necessary to carry out preliminary repair works to a structure to prevent damage by the cleaning method. For example, where cracks are wide enough to let moisture penetrate deep into the concrete, water may exacerbate corrosion of the reinforcement. Other protection works to openings and so on may also be required.

Types of soiling and cleaning methods

Green algae and fungus

In the UK there has been a significant shift from the use of coal to natural gas for heating and power generation over the past two decades or so. As well as reducing black pollution staining, the resultant drop in acidic gas pollution in the atmosphere has probably caused an increase in organic growths on buildings. Essentially, the pre-pollution natural colonisation and subsequent decay of materials have returned to some urban environments. Buildings may go green instead of black. This will occur particularly towards the ground or around any damp areas. High-pressure water lances are probably the easiest method of cleaning. Brushing the algae may smudge it into the surface rather than wash it away, and then necessitate the use of large amounts of water to rinse down the surface. The lowest water pressure compatible with effective cleaning should be used, as the jetting may cause damage to the surface of the concrete.

If a biocide is used to stop the growths from returning, it may be possible to kill the algae first to make cleaning easier. This may require some time. Biocides have only a finite lifetime and need to be re-applied. They can have wider environmental impacts if the residues are not controlled or disposed of correctly.

Lichens

Lichens prefer sunny areas on buildings and structures which are well drained, but still receive water. On concrete buildings they tend to populate the vertical and horizontal surfaces that are fairly exposed. They often produce a dark grey overall appearance that may look like pollution soiling, but in fact is a natural organic growth. Sometimes there may be pollution soiling of the aggregate, mixed with lichen in-between, over the cement matrix. This will require other cleaning methods, detailed below for pollution soiling. Lichens may become much easier to remove when they are wet. Thus a pre-soaking of the surface is recommended, and the cleaning must be carried out when it is still wet. Water washing with brushes is probably the easiest and least damaging method of removal. The optimum soaking time for easiest removal should be determined during trials.

Pollution soiling

Black crusts from environmental pollution can adhere very strongly to concrete (Fig. 6.2). Sometimes it is present on the cement matrix, or it may just be present on the aggregate particles. Care is required to ensure that damage is not done to the concrete surface, the aggregates or the interstitial cement matrix as separate surface components. There follows a description of a number of possible approaches to cleaning pollution-contaminated surfaces.

Water washing combined with stiff brushing and scrubbing

This can be an effective method for removing pollution soiling, but it can also be slow and produce an uneven effect. The effect of stiff brushes can be physically damaging to the concrete surface and can also damage moulded finishes, and softer brushes should be used for decorative finishes.

Steam cleaning

There is a wide variety of commercial apparatus for steam cleaning, with very different volume deliveries and steam pressures. The smallest may be considered 'micro' steam cleaning for conservation projects. The largest may be similar to high-pressure water lances for general cleaning. Care is required with moulded finishes.

Fig. 6.2   Great Western Goods Shed, Bristol (1910). Note the soiling under the window ledges from environmental pollution in areas that are not regularly washed with water. Trials were an important part of the project during the writing of the specification. The figure shows trial external finishes (photo Susan Macdonald).

High-pressure water lances

These can be very effective for the removal of loose or softened dirt and residues, especially when combined with other cleaning methods. However, used incorrectly, water lances and jets can cause substantial damage to all concrete finishes. Even on exposed aggregate surfaces, water jetting may selectively remove certain types of aggregates or etch the interstitial cement matrix. This may change the general appearance and exacerbate deterioration.

Used at full power and very close, the water jets can reach supersonic speeds. The effect is physical shock waves to the concrete surface. If there are any pores, cracks or holes that can fill with water, a hydraulic effect may sweep away aggregates or fittings, or extend cracks. This can occur deep into the material. The pressure of some water jets used for the demolition of concrete structures (15000 p.s.i) is only ten times higher than those commonly available with water lances or water jet cleaning systems (1500 p.s.i).

Water lances, either on their own or combined with other cleaning methods, require careful control of the water pressure, flow rate and nozzle-to-surface distance. They can also be hazardous if used at either high water pressures or high temperatures. There are clearly health and safety implications in the use of this type of equipment that need to be taken into account when planning their use on a particular project.

Chemical cleaning

Generally the chemicals, whether acid, alkali or detergent, are in the form of a gel or poultice that allows easy application. The gel or poultice supports the chemical on the surface for the required dwell time or cleaning agitation period. Chemicals tend to break down the black soiling or the concrete surface it is adhering to, and so provide an easier method of removal by brushes or water washing later. Water lances are often used to remove the residues, chemicals, gels and poultices. However, residues from the chemicals may still contaminate the building and cause efflorescence. Chemicals can often 'over clean' and give a bleached appearance.

The cleaning process requires extreme care where water lances are used in conjunction with other cleaning methods to remove cleaning chemicals and residues. The water lances have their own cleaning effect. It may be unclear to what extent the cleaning was produced by the chemicals or by the abrasive action of the water jetting. Trial cleaning is generally the preferred method of determining the best approach, and to ensure that an even result can be obtained.

There is a wide range of mixed chemical products now available, making generic chemical classification difficult. Therefore, it is not useful to list the types of chemical used in cleaning products here. For example, acids can be mixed with detergents, detergents can be mixed with alkalis, and alkalis can be mixed with solvents. In addition, their effectiveness is often related to the design of the thickeners (gels and poultices) and the workers' skill when using the cleaning product. An individual commercial cleaning product may be the most appropriate choice, not because it is generically acid A or acid B, but because the contractor concerned is skilled in its use. Chemical cleaning products should perhaps be considered as a tool rather than a treatment.

The following points should be considered when adopting a chemical cleaning method:

- hazard data sheets (which must be supplied with all chemicals) should provide a clear identification of the chemicals in the cleaning agent;
- it can be very difficult to determine what type of cleaning chemical is best without trying several with different chemical contents;

- agitation with brushes, increased 'dwell time' and re-application may dramatically increase the effectiveness of chemical cleaning;
- chemicals, agitation with brushes, increased dwell times and re-application may cause bleaching and surface damage.

Abrasive cleaning

There is a wide range of abrasive cleaning methods. It is not possible to classify them clearly and simply. Terms such as 'abrasive cleaning', 'grit blasting', 'soft air abrasives', 'grit slurry', 'sand blasting' and 'micro-air abrasives' are used to describe the processes. Companies and suppliers of such systems may produce marketing material and brands designed to amplify the differences between them. The aim is to persuade speci-fiers to consider such systems separately. However, from the point of view of assessing the effectiveness of abrasive cleaning, they all work in essentially the same way.

Abrasive particles are blasted out of a nozzle using pressurised air. When they hit the building surface, they fracture the soiling or concrete surface and so effect cleaning. Some systems have water entrained into the system. This reduces dust problems and can actually improve the general abrasive cleaning affect. Some systems vary the abrasive parti-cles. The so-called 'soft' abrasives use nutshells or weaker carbonate and silica materials, or similar, as the abrasive particles. The idea is that if the abrasive is softer than the concrete, it cannot cause damage. However, it is difficult to determine the impact of such materials when they are fired using pressurised air. The shock waves caused by the impact of these particles can still cause fractures in harder materials.

The size and scale of the apparatus is a good guide to the possible speed of cleaning and the controllability of cleaning and damage. Small-scale micro-systems can be good for the removal of graffiti and stains, and for cleaning more delicate mouldings and detailed surfaces. Large-scale equipment can clean large areas extremely quickly. However, the fast cleaning methods may also produce more damage. With larger-scale systems where water is entrained in the system, substantial amounts of spray can be produced. This obstructs the view of the person cleaning, and so mars their ability to judge the cleaning effect achieved. The dust produced by dry systems can be hazardous to health.

Localised stains and graffiti

Stains and graffiti are unsightly and can greatly affect the appearance of the building. Their localised nature raises particular cleaning issues. Some

Fig. 6.3 St John and St Mary Magdalene, Goldthorpe, South Yorkshire, 1914–1916. Rust staining of the concrete below the windows due to the corrosion of the metal window guards (photo Susan Macdonald).

staining may be related to rust from embedded metal reinforcing bars (Fig. 6.3). When 'tough' stains are cleaned, they can leave an excessively clean patch or 'ghost mark'. To remove stains and graffiti effectively from prominent areas without 'ghosting' requires a great deal of care and often small-scale testing. Solvents containing chemicals or abrasives may be most effective in graffiti removal. It may be necessary to experiment with chemical or abrasive cleaning methods. If the building is historic or prestigious, then it is often best to consult experts in building conservation and repair; listed buildings may require statutory consent prior to treatment.

Salt accumulation and lime-blooms

Dilute hydrochloric acid (ca. 5%) will remove salt accumulations and lime-blooms from concrete surfaces. Thick encrustations of lime may require mechanical removal with tools or abrasives.

Cleaning of concrete repairs, sealants, joints and fittings

Care is needed where cleaning reveals, or takes place, over joints and other fittings, or materials embedded within or close to the concrete surface. Concrete does not easily adhere to other construction materials after it has set. As a result, any method that removes soiling may also cause disruption of the bond between the concrete and other building components.

Concrete repairs may be particularly susceptible to disruption by high-pressure water jetting. Metal fittings may decay with the use of acidic chemicals. Solvents may cause deterioration of sealants.

Just blitz it?

If there is really no need to consider the long-term appearance or the integrity of the concrete surface, then high-pressure water jetting, or sand blasting, will generally be able to remove any soiling. However, such methods are capable of removing several millimetres of the surface if used aggressively. Such an approach may expose the larger aggregate particles. These aggregate particles may be a completely different colour, or even very dark. The result could be a greater change in appearance than envisaged. In addition, reinforcing bars may be only a few centimetres below the surface. Erosion of the surface may reduce the concrete cover over the embedded metal, possibly causing some degree of reduction in the future durability of the concrete structure concerned.

Management of surface cleaning projects

It is generally advisable to seek specialist advice for a cleaning project, especially where it is technically demanding, where high-quality results are required or where there are tight project time-scales. The specialist should also be able to advise on project management issues.

Without a specialist, the owner/manager will be relying on the expertise and knowledge of the main contractor in respect of the method(s) adopted for surface cleaning. Care is needed with this approach. Even where the building itself is of little importance in terms of conservation or in the quality of the cleaning that is to be achieved, there are still important project management and technical considerations. It is self-evident that the lower the technical requirements for a surface cleaning job, or where there is less need to achieve good-quality work, the lower the likely skill level of the workers employed.

It is worth noting that main contractors or 'cleaning system' providers may subcontract local firms to actually do the work. Whilst this may result in a lower 'first-cost', there are risks associated with such savings. The local firms may employ only semi-skilled or unskilled workers. Such issues are discussed more generally in relation to the management of maintenance works projects in the section on management of maintenance works projects later in this chapter.

The following points are often relevant where the lowest skill level is to be adopted for a cleaning project.

1. It should be clear that trial cleaning should be carried out either in advance of the contract, as a separate contract, or as part of the specification. All contractors involved in the cleaning should provide written method statements for what they are going to do.
2. In negotiating the cleaning method, it should be noted that water lances are the cheapest possible cleaning method. They require no chemicals to dispose of, no heavy lifting of apparatus, and no estimation of material use. The abrasive, water, is almost free. As a result, if the contractor has any choice in the matter, then they will tend to select water lances. Excessive wetting of a building may well have implications for the durability and performance of its surface finishes, especially of masonry. Pre-cleaning repair works may be necessary to protect against water ingress.
3. If a cleaning contractor is using specialised chemical or abrasive cleaning to a specification, they may justifiably like to include water lances as additional assistance for 'washing down'. However, if unscrupulous contractors are short of time, or using too much expensive chemical cleaner, they may increase the pressure of water-jetting used in rinsing to assist in cleaning. They may bring to the site a much bigger water-jetting apparatus than is strictly required. This is likely to cause damage to the building. Restricting the types of equipment brought onto the site, and regular inspections by project managers, may be the only way to control the inappropriate use of high-pressure water jetting.

Health and safety

Cleaning is potentially a dangerous process, and individual responsibilities should be carefully determined. Construction Design and Management (CDM) regulations 1994, the Health and Safety at Work Act 1974, the Management of Health and Safety at Work Regulations 1992 and the Control of Substances Hazardous to Health Regulations (COSHH) 1999 are all relevant to building cleaning. The roles and responsibilities according to the law are changing, and now incorporate the project design and the designers (CDM).

It may be appropriate to question the main contractors and their subcontractors (where relevant) about their approach to meeting their legal responsibilities. Areas of questioning might include:

- what are the procedures in the event of a spillage of a chemical cleaning product?
- where are the hazard data sheets for the cleaning chemicals being used?
- where are the chemical containers and are they all labelled with their contents?
- do workers ever have to remove their protective eye goggles while they are cleaning to see what they are doing?
- what are the procedures in the event of a physical injury?
- when did you do this type of cleaning before?
- how will the residues from the cleaning be disposed of safely?

Final visual assessment

The lighting conditions directly affect the appearance of the final cleaned building. The cleaned areas viewed in trials may not produce the same effect when they are extrapolated to the rest of the building. In addition, viewing in wet, cloudy or sunny conditions can also change the appearance dramatically.

Coatings for concrete

Introduction to surface coatings

Basic principles

Surface coatings comprise an extremely diverse range of product types designed to decorate and/or protect substrates to which they are applied.

They are often known generically as *paints*, although strictly this defines a product that contains colouring materials and is opaque. Clear coatings such as varnishes, lacquers and water-repellents are also produced and can be considered to be non-pigmented *paints*.

Composition of surface coatings

Coatings can be considered as being composed of a polymeric binder (resin) that is normally dissolved or dispersed in a liquid carrier phase (solvent); a paint coating will also contain finely dispersed particulate materials (pigments). These basic constituents would form the bulk of the composition, although all coatings contain minor amounts of additives to control and modify the properties of the final product.

Binders

The binders used in coatings are polymeric complexes that provide integrity to the dried film and bond it to the surface to which it is applied; binders are often referred to as resins or polymers. A wide variety of binders can be used in the manufacture of coatings and, with some exceptions, these are all organic in composition. Binders can be divided into two broad groups of convertible and non-convertible types. Convertible binders are materials that are used in an unpolymerised or partially polymerised state and which, following application to a substrate, undergo a reaction to form a polymerised (solid) film. Non-convertible binders are pre-polymerised materials that are dispersed or dissolved in a carrier solvent; following application, the carrier evaporates to leave a coherent film on the substrate surface.

The main binder types that could potentially be used in coatings for cement-based substrates are listed below.

1. Convertible coatings:
   - oils;
   - alkyd resins;
   - epoxy resins;
   - polyurethane resins;
   - silicone resins.
2. Non-convertible coatings:
   - chlorinated rubber;
   - vinyl resins;
   - acrylic resins.
3. Miscellaneous binders. Two other binders are important in paint technology, although they are not strictly classified as polymers:

- bitumens (USA, asphalts);
- inorganic silicates.

Pigments

A pigment can be defined as a solid material in the form of small discrete particles that remain insoluble in the resin and solvent constituents of the coating. Pigments are added to impart colour and opacity, and/or modify the protective efficacy of the coating on the substrate. There is a further class of pigment that whilst modifying the properties of the coating by, for example, supplying water resistance, confers little or no colour or opacity; these are called extenders. Extenders are inorganic in origin, but pigments can be either organic or inorganic. Coatings for cementitious substrates may contain pigments and extenders, although the pigments are most likely to be of the inorganic type owing to their greater durability and resistance to chemicals.

Solvents

Solvents are low-viscosity liquids, often referred to as volatile organic compounds (VOCs). Water is an important solvent in many modern coatings, although it is not considered to be a VOC. The function of solvents is to carry or dissolve the resin component, provide appropriate conditions for pigment dispersion, stabilise the finished product during storage, control application characteristics and aid film formation after application. In some coatings, solvents may be the major component of the final product, although coatings with a low solvent content have increasingly been made available to users.

The solvents most widely used in coatings are listed below.

1. Hydrocarbon solvents:
   - toluene;
   - white spirit;
   - xylene.
2. Alcohols and ethers:
   - butyl alcohol;
   - ethyl alcohol;
   - ethylene glycol monoethyl ether.
3. Esters and ketones:
   - acetone;
   - methyl ethyl ketone;

- methyl isobutyl ketone;
- butyl acetate;
- ethyl acetate.

The solvents described above find uses across the whole field of coatings technology. However, only a limited number find applications in the types of surface coating used on concrete construction. These coating types are identified in the next section.

It should be noted that there are increasing health and safety and environmental concerns about emissions from all types of VOCs. Health and safety concerns centre around issues relating to flammability, storage and transportation of solvent-based coatings, and the protection of individuals in occupations where coatings are produced and used, or in the immediate vicinity of such operations. Environmental concerns over solvents have proved more difficult to address and constitute greater cause for concern. It is now recognised that the organic solvents present in almost all surface coatings are major pollutants. They react with other emissions, notably those from car exhaust fumes, to create ozone that, at ground level, causes photochemical smog which is implicated in crop and tree damage. Photochemical smog is also believed to have an adverse effect on individuals suffering from respiratory illnesses such as asthma.

Additives

Many additives are used in coatings in order to control or modify properties such as viscosity, flow and the drying time of the coating, and to inhibit biological growths. These additives are generally used in low concentrations relative to the main formulation ingredients described above. The uses of some of the important types of additives are:

- plasticisers;
- driers;
- bioactive additives;
- viscosity modifiers and anti-settling agents;
- pigment-dispersing agents.

Selecting a suitable coating type

Whilst some cement-based surfaces are inherently weatherproof, durable and aesthetically acceptable, others are painted for a variety of reasons: to provide colour or surface texture, to provide protection from the environment, to reflect or absorb light, to hide repairs and/or to facilitate cleaning, or for hygiene purposes.

The durability of a coating system is its ability to provide decoration and/or protection to its substrate over the period of time known as its effective service life. This period is the interval between the time of application of the paint system and the moment at which, through deterioration, it ceases to perform its required function. The durability of paint films is dependent upon a number of factors. These can be divided into 'internal' factors, such as formulation details and their effects on the physico-chemical characteristics of the coating, and 'external' factors, such as the nature of the substrate, substrate preparation, and environmental effects such as location and weather.

Types and characteristics of coatings

Details of the principal ingredients available to coatings manufacturers have already been discussed; aspects of combining these into coatings with the requisite properties for use on cement-based substrates are now described.

Coatings for interior surfaces

Interior cementitious surfaces are coated primarily for decoration. The following paint types are those most commonly used for interior surfaces:

- water-borne emulsion paints;
- alkyd resin-based paints.

Emulsion paints are based on a variety of polymer binders; their general properties are dependent on many factors other than the nature of the binder. Because they are known to be permeable, and can be used when walls are not completely dry, they are often misused when applied within days of casting. Although relatively trouble-free, emulsion paints occasionally give rise to difficulties with efflorescence or patchiness on walls, in which case a penetrating sealer must be used first. A flat oil paint or a sealer, followed by more emulsion paint, can be used to make good an existing faulty paint film.

Better general dirt and wear resistance can be achieved with oil-based (e.g. alkyd) matt finishes, but these are now rarely used; they require more skill in application and need an alkali-resistant primer on most new surfaces. They are also more likely to yellow with age than emulsion paints. Types that are claimed to allow brush-cleaning in water have recently been introduced.

The solvent used in alkyd resin-based paints is the aliphatic hydrocarbon, white spirit. It should be noted that water-borne variants have recently been introduced with similar properties to alkyd-based paints, and these are preferred for reasons of health and safety.

Coatings for exterior surfaces

Exterior surfaces are generally painted to obtain the required decorative effect, although in some applications, such as the use of anti-carbonation coatings on concrete, the paint may provide an additional function. The most commonly used types are:

- water-borne emulsion paints;
- solvent-borne finishes;
- cement-based paints;
- alkali silicate-based (mineral) masonry coatings;
- textured finish paints.

Exterior emulsion paints are generally applied coat-on-coat to build up the required thickness on the surface. On aged surfaces that are weak and friable, or if there is excessive suction in the substrate, then primers may be used before the application of the finishing paint. For exterior masonry, these primers are designed to penetrate deeply, and this requires that they be based on solvent-soluble resin systems. To further aid penetration, a relatively low solution viscosity is necessary, so solvent contents are often very high. Aliphatic and aromatic hydrocarbon solvents are used in these primers.

The solvent-borne finishes have certain performance advantages over the water-borne emulsion paints, for example, greater tolerance to adverse weather/substrate conditions during application. The resin type typically used in these paints requires the use of aliphatic and aromatic hydrocarbon solvents to effect solubilisation and provide the requisite coating properties.

The painting of new cement-containing substrates can prove difficult, and problems may arise using the coatings described above. Traditionally, cement paints have been considered most suitable for very early decoration and they can be applied within a few days of the substrate setting. Cement paints are based largely on Portland cement with specific additives to control application, setting time and colour, and the tolerance of these paints to fresh cement-based surfaces lies in the similarity between the paint itself and the substrate.

Cement paints are supplied in powder form and are mixed with water to a suitable consistency prior to use. Application is generally two coats

Fig. 6.4   House at Sydenham Hill (architects Harding and Tecton, 1935) after repairs and repainting. The house was originally finished externally with 'Castex', a spray-applied cement slurry product. The concrete walls were originally cream, with eau-de-nil soffits and recesses (photo Susan Macdonald).

by brush, allowing 16–24 hours between coats (Fig. 6.4). The level of pigmentation is critical in cement paints, and generally a figure of 5% is not exceeded; additions of pigment above this level can result in reduced film integrity in the resultant coating. However, these relatively low levels of pigment are adequate to produce a wide range of colours, and where dark shades are required, the use of ordinary Portland cement that is grey in colour is normal. The use of pigments is restricted to inorganic types with high alkali resistance. The other additives in the formulation modify the application rate and film properties. Stearate addition improves the application characteristics of the paint and upgrades the water-resistance of the film, whilst the addition of calcium chloride regulates the drying time of the applied coating. Hydrated lime can also be included in order to reduce the cracking tendencies associated with the hardening of the cement component.

Alkali silicate-based (mineral) masonry coatings, which are becoming increasingly popular in the UK, have had a long history throughout much of the rest of Europe. They are based on an alkali silicate solution (usually sodium or potassium silicate) in water, and are pigmented with alkali-resistant pigments. They share similar characteristics with cement paints, but are considered to be far superior in durability and may be used on cement-based and other mineral substrates in preference to masonry coatings based on organic binders. Alkali silicate-based masonry coatings may be modified by additions of aqueous polymer dispersions, usually an acrylic co-polymer having good resistance to alkali, in order to modify their drying characteristics under adverse conditions, and other physical properties.

Textured finish paints are the most widely used exterior wall coatings and cover a wide range from very fine sand or stone textures to those containing coarse aggregate or worked into patterns by rollers or hand tools. There is still some use of the old established stone-textured paints based on oleoresinous media, with sand or mica fillers, which have a useful life of around 7 years, giving a reasonably economical balance between first cost and maintenance. About the same life is obtained from the more popular exterior emulsion-based paints of similar texture.

Thick, textured, sprayed coatings (about 0.5–0.8 mm), usually based on 'polyester' (i.e. alkyd) resins with mica, perlite and sometimes fibres, have a life expectancy of over 10 years if correctly applied. Some are described rather incorrectly as 'plastic'. There are also similar types of coating based on oil-free polymers, chlorinated rubber or emulsions. Most sprayed textured coatings are offered with a 'guarantee' by specialist application contractors. This, although limited to the behaviour of the coating in specified ways, can be a useful feature if the operators or manufacturers are known to be of good standing and with several years background of successful applications. Nevertheless, at best, these expensive treatments may not show the lowest overall 'cost in use' over a long period on domestic buildings, for which the most economical treatment is probably with masonry paints at about half the cost of the textured coating and half the life. The longer-life coatings have useful advantages on buildings where application costs are high, or there is a requirement for a high standard of appearance or better resistance to rain penetration, for which extra cost can be justified. Also, the possibility of eventually recoating the original textured coating with a thinner masonry paint would lower the calculated cost in use.

The thick, textured coatings can also contribute to the rain resistance of an external wall, but are too thin to substantiate claims for acoustic and thermal insulation properties except that by reducing the moisture content of the wall they would prevent the reduction of its original thermal insulation value.

Fine- and coarse-textured coatings applied by roller produce films 1–3 mm thick, at application rates of 0.8 to 2 or 3 kg/m², and can be given a variety of textures by the method of working. To date, these have been emulsion-based, but solvent-thinned systems are appearing. Within the rest of Europe, they have been described as organic or plastic renderings. Although expensive in materials, they are low in labour costs. The durability of (water-thinned) examples has been disappointing in Building Research Establishment (BRE) tests with excessive dirt retention and susceptibility to damp conditions, e.g. near the foot of walls. Given good rain-resistance, they would be able to save the cost of a preliminary render coat, but even at a thickness of 1–2 mm they can barely hide joints in blockwork, and those with a scratched or scraped effect cannot be resistant to rain penetration. Other forms of emulsion-based thick coatings, applied by spray or trowel, contain mineral or glass particles in a clear binder with no pigment, giving a translucent or multi-coloured effect. These also have a life expectancy of over 10 years, with similar reservations as to dirt pick-up (less important with the darker colours).

Primers and sealers

Many paints are applied over primers or sealers that may modify the properties, especially the permeability, of the finish. Although it is generally desirable to use sealers as directed by the manufacturer, some examples examined at BRE have proved to be unsuitable or even incompatible with their finishes. In particular, emulsion-based 'sealers' do not penetrate well into surfaces of low permeability and may lead to adhesion failures of coatings. Even solvent-thinned sealers may not penetrate and bind down multiple coats of old paint, as is often required, and they should not be used on sound surfaces of low or moderate porosity. Therefore, it is important to consider primers or sealers as an integral part of the coating system.

Special performance coatings

There are a number of coating types that perform special functions in addition to decorative purposes, and these are discussed in the following section.

Surface water-proofing treatments. Where rain penetration has been established as the cause of internal dampness (rather than condensation or rising damp), and where the appearance is required to be maintained and not hidden by paint, colourless water-repellent treatments may be used. These include silicones and aluminium stearate compositions and acrylic resin solutions. Much of the rain penetration of blockwork walls occurs via fine cracks in the mortar joints, and water repellents will not always seal these. Water shed from treated areas tends to run into cracks and may then make localised dampness worse. Treatments should be applied only after cracks and defects have been made good: repointing may even be necessary.

Colourless treatments should be applied at adequate strength and possibly as a double application. They cause some retardation of drying out (but less than most paints) and, if crystallisable salts are present, may result in spalling of the surface. Their life can be 5–10 years, but renewal is desirable when their visible water-repellent properties are lost.

Clear solutions of acrylic polymers can be more effective where there are fine cracks, but they tend to produce more of a film on the surface, and hence a glossy appearance with darkening of the colour of the substrate. They are often used on concrete and exposed aggregate concrete to reduce rain penetration. They are more restrictive of evaporation from the surface and their durability is less than that of paints, but they may possibly continue to be effective for up to 5 years. In comparing available products, low viscosity at high solids content (10–15%) is desirable to ensure good penetration, although this will be slight into dense concrete. Clear polyurethane waterproofing compounds have not shown satisfactory durability.

Pigmented coatings form a better barrier to rain and in general have a longer life. The thicker masonry coatings, and especially the sprayed textured type, are reasonably effective although some mineral-filled emulsion coatings may shed light rain but allow penetration by driving rain over long periods, as do water-repellents. One popular system utilises a bituminous emulsion-waterproofing coat, followed by a specially formulated decorative exterior emulsion paint. Another system uses a pigmented rubber polymer, producing thick smooth decorative coatings (sometimes described as liquid plastics) with a life expectancy of over 10 years on walls, and has good crack-sealing properties until the coating eventually becomes brittle. Some treatments include a reinforcing glass or synthetic fibre cloth to improve performance at cracks and joints.

Attempts to seal walls on the inside are usually less successful than external treatments, but the various proprietary compositions sold for this purpose can be useful against moderate or intermittent dampness. Some are based on polyurethane formed by reaction with moisture in the wall, but the amount of moisture they take up is very small. Their performance has not been studied by BRE. If moisture in the wall cannot escape by another route, impervious coatings are liable to be disrupted. Possibly their most useful application is where the dampness has brought out stains, on chimneybreasts for example, which would affect other paints and have to be sealed.

Walls below ground should be sealed on the outside before backfilling; bituminous compounds and emulsions are effective, economic and durable. On the inside face, most coatings are unsuitable when water has already penetrated, although some proprietary treatments using rapid-setting cements in conjunction with overall coatings are claimed to be effective.

Anti-graffiti treatments.   It is virtually impossible to remove all graffiti types completely from a porous substrate such as concrete without spoiling its appearance. In some instances, graffiti and spills may need to be painted over, and some manufacturers will supply paints specially matched to substrate colours. Anti-graffiti barriers provide protection for areas where graffiti is an on-going problem.

Two main types of barrier are used for protection against graffiti, the so-called 'permanent' graffiti barriers and 'sacrificial' graffiti barriers. Permanent barriers seal the surface and form a coherent coating that is not readily penetrated by graffiti in the form of paint or ink. The graffiti can therefore be removed with cleaning agents, leaving the protective coat intact. Depending on the type and severity of the graffiti, a two-stage approach to removal may be required. This comprises initial cleaning with solvents or detergents and then the use of poultices to remove any residual staining within the coating. Permanent graffiti barriers are produced by two-pack polyurethane coatings, supplied specifically for this purpose by several manufacturers.

Sacrificial graffiti barriers are based on organic compounds and are removed after an attack, along with the graffiti, and then reapplied.

Both types of anti-graffiti treatments are normally supplied in unpigmented (clear) form with the intention of leaving the appearance of the surface largely unaffected. However, treated surfaces often exhibit slight colour change and may also develop a glossy sheen. The service life of anti-graffiti treatments will depend upon their exact composition, but some treatments are claimed to remain resistant to graffiti for up to 10 years without re-treatment.

The application of anti-graffiti coatings generally requires no special skills and can be undertaken by professional painters. However, removal of graffiti, especially from large areas, relies on correct diagnosis of the marker and an understanding of the substrate, and often requires the use of aggressive chemicals and processes. Therefore, it should be undertaken by specialist contractors.

Many suppliers of anti-graffiti treatments can offer detailed advice on specification based on experience developed over many years. This area of coatings technology is extremely specialised, and though the general advice given here outlines the basic principles of the use of anti-graffiti treatments, detailed recommendations should be sought from manufacturers if the use of such products is being considered. Research by English

Heritage has shown that the sacrificial anti-graffiti treatments can introduce a new set of problems for historic masonry, such as on-going effectiveness, substrate durability and changes in appearance (English Heritage, 1999). For these reasons both permanent and sacrificial graffiti barriers are unlikely to be appropriate for listed masonry buildings. Likewise, they may not be suitable for concrete buildings with delicate or weathered finishes where the barrier is likely to have a visual effect: the repeated removal of sacrificial graffiti barriers may also be damaging.

Anti-carbonation coatings. Carbonation occurs when carbon dioxide in the air penetrates the pores of the concrete and reacts with calcium hydroxide to form calcium carbonate and water. This process reduces the alkalinity of the concrete, and removes the protection afforded by the concrete matrix to any embedded reinforcing steel in the surface zone affected, potentially allowing corrosion to occur. The risk and extent of corrosion may be increased by the presence of chloride in the concrete and by the severity of the environment. This process is discussed in more detail in Chapter 4. Where deterioration is present, an investigation will usually be required to identify the causes of deterioration and the areas in need of repair. This will allow a suitable repair procedure to be selected. The processes of deterioration are described in detail in Chapter 4, with options for repairing damaged concrete being outlined in Chapter 5. For aesthetic reasons and to minimise the risk of subsequent deterioration, the application of a coating often forms part of the repair procedure.

If carbon dioxide can penetrate the concrete matrix, then oxygen can also. The factor controlling corrosion will therefore be the availability of moisture at the steel surface. A coating may be used to help control this.

A coating may be applied over the whole exterior surface of the concrete to reduce the rate of carbonation by acting as a barrier to the penetration of carbon dioxide. It should also prevent the ingress of liquid water, but allow water vapour to pass through it. This will prevent entrapment of liquid water in the concrete and between the coating and the substrate. The coating treatment must, of course, be suitable for application to concrete. Furthermore, for coatings to serve as a barrier to carbon dioxide they must form a defect-free film. The surface of concrete is irregular and uneven; blowholes and other surface imperfections are quite common, so multiple coats are required to reduce the risk of coincident defects. The performance of coatings diminishes with ageing, in some cases to a marked degree.

Work carried out at the BRE indicated that for coatings based on acrylics and chlorinated rubbers, there is a correlation between coating thickness and permeability. Permeability clearly decreases with greater coating thickness. The results indicate that barrier coatings should be applied to a total mean thickness of at least 200 μm to achieve adequate film thickness over the whole surface. If the concrete to be coated is very rough or irregular, then consideration should be given to making the surface level and uniform by means of a 'fairing' coat.

Similarly, when the concrete is very porous, a sealer of a penetrating resinous type can improve substrate uniformity.

Systems that consist solely of clear coating materials have little resistance to carbon dioxide penetration because they cannot be applied thickly enough. Thus, whilst such materials may be desirable for aesthetic purposes, they do not adequately reduce further carbonation. The coating system should therefore include a pigmented topcoat. Experience has also shown that clear coatings are not as durable as pigmented coatings.

It should be noted that coatings may only be used as part of the process for the repair of concrete where the risk of corrosion due to chlorides is low. Where there is a greater risk, then the use of surface coatings without any remedial action is not generally considered to be appropriate.

Impregnation treatments for cement-based substrates. The coatings described above have been film-forming, i.e. intended to deposit a layer of appreciable and measurable thickness on the surface. Most coatings for concrete protect simply by providing a durable and resilient barrier between the substrate and the external environment.

There are some products however, that whilst classed as 'coatings', protect by being absorbed into the outermost layer of the substrate, where they confer some extra property. Strictly speaking these are impregnation treatments rather than coatings, but since the underlying technologies are similar, a summary of them is appropriate here for completeness.

The most common use for impregnation treatments is to enhance the water-repellence of the surface. The properties and composition of impregnation treatments are defined by the key requirements for complete absorption and high penetration. Treatments are therefore simple low-viscosity solutions of appropriate resins or resin combinations in a solvent, but without any pigment addition that might interfere with penetration. Therefore, once impregnation treatments are applied, the appearance of the surface remains largely unaltered.

Various derivatives of organic silicone resins, such as silanes and siloxanes, have been used for many years to improve the water repellence of cement-based substrates. In particular, they have been widely used to increase the water-repellent characteristics of reinforced concrete, because water (and any aggressive chemical species it may contain) plays a key role in the corrosion of the steel reinforcement. Impregnation treatments function by lining the pores of the concrete, thereby significantly reducing the capillary absorption capacity of the substrate and lowering the amount of water that can be absorbed. An advantage of impregnation treatments is that whilst they inhibit liquid water absorption, they have little effect on the diffusion of water vapour, thereby allowing the moisture content of the substrate to equilibrate with that of its environment.

The Highways Agency has experience of the use of silicone-based impregnation treatments. BD 43/90 (1990) provides the criteria and specification for a silane impregnation treatment for the protection of concrete highway structures against reinforcement protection.

Substrate considerations

The nature of the substrate and its preparation prior to coating will have a major effect on the service life of a coating system. The main concrete substrate types and their implications for a selection of coatings are described below.

Portland cement rendering

The alkalinity of Portland cement is, with rare exceptions, so high that precautions against alkali attack of paints by new cement-based products should always be taken. Because of the caustic alkali present, carbonation of the lime content at the time of painting is not always sufficient to prevent the saponification of oil-paints. Even when the surface pH falls to about 9, there may be further alkali at a greater depth that will affect paints as moisture moves through the substrate. The free lime which is usually present in cement renders is a cause of the lime-bloom that is particularly noticeable on emulsion paints. If carbonation is allowed to occur before painting it can greatly reduce this effect, but it may also reduce surface porosity and produce shrinkage crazing.

Cement and alkali silicate-based paints are suitable for very early decoration; acrylic emulsion or other alkali-resistant porous paints can be used after about 4 weeks drying. For a glossy finish, the moisture content must be down to a safe level and several months drying may be necessary. An alkali-resistant primer is needed under oil paints; hence non-saponifiable (e.g. chlorinated rubber) paints are preferable. Chemically resistant wall finishes require a hard base: cement rendering is normally suitable if it is dry, but if a very smooth finish is needed for internal walls, an anhydrous plaster can be used on top.

It may be thought that the craze patterns and eventual cracks that develop on painted rendering are unavoidable and not a fault of the paint. However, many comparative tests at the BRE have shown that paints differ greatly in their ability to retard this crazing. Whilst this effect may appear within 1 or 2 years under thin emulsion or masonry paints, it may not occur for many years where the coating is thick, flexible and not highly permeable to moisture and carbon dioxide.

Concrete

Concrete is considered to be a durable, permanent material, but long-term deterioration mechanisms may require protection to be applied to the surface. Following deterioration and repairs, or even when new in the absence of sufficient cover to protect the embedded steel reinforcement, there is often a requirement to provide protection against penetration by water, carbon dioxide, sulphur dioxide and salts. There is also an increasing need for decorative or colouring treatments to improve the appearance of large areas of concrete, or even for colourless waterproofing treatments to prevent the accumulation of dirt and biological growths.

The general requirements for painting concrete are similar to those for cement renderings, but the surface may be either hard and smooth with poor adhesion for paints (e.g. high-quality pre-cast components), or rough and porous with relatively large surface voids and blowholes. Residues of mould oil can seriously reduce paint adhesion; they are best removed by abrasion rather than solvents, but detergents or emulsifying agents may be useful. The application of cement-based paints, or 'bagging' with a cement–sand mix, will help to fill voids and blowholes. Larger defects may be filled with mortar or epoxy-resin mortars, or, internally in dry conditions only, with gypsum plaster or water-mixed powder fillers. Alternatively, thin-wall plasters (internal) or textured paints (either internal or external) may be used to hide surface blemishes.

Where concrete cover is known to be low, no firm statement can be made on the equivalence of coatings to a desired thickness of concrete. For reducing the ingress of water to concrete where corrosion of reinforcement is considered likely or has been made good, impervious coatings based on chlorinated rubber, two-pack epoxy resins or polyurethanes are suitable provided the concrete is fairly dry, with a potential life of 7–10 years to first maintenance. Their benefit will be more doubtful where chlorides are the cause of corrosion. Coatings which are less impervious, to allow drying-out but with a claimed degree of resistance to the passage of carbon dioxide to minimise carbonation of the concrete, are now being marketed. The principle appears valid, but more practical evidence is required.

Concrete blocks and calcium silicate bricks

Dense concrete blocks and calcium silicate bricks may be painted as for concrete. Some light-weight blocks allow rain penetration and may need complete protection (e.g. a thick bituminous coating, with decorative coats if needed), or merely rain-shedding and vapour-permeable protection (e.g. thick textured emulsion paints).

Aerated (and lightweight) concrete requires paints with some filling action; emulsion paints may suffice internally and for purely decorative purposes, but if resistance to rain penetration is a requirement, especially in exposed situations, thick textured relatively impervious coatings may be necessary, with the alternative of a rendering. These will delay drying out and should therefore only be applied when the moisture content of the aerated concrete is at a suitably safe level.

Environmental considerations

The environmental conditions that paint films are exposed to during service are of great importance in determining their performance. Knowledge of the anticipated service conditions is essential when formulating or specifying paint coatings. Environmental factors can be subdivided into the effects of location, i.e. atmospheric pollution, and the effects of weather.

Influence of location and atmospheric pollution on coating durability

British Standard BS 6150 (1991) identifies four categories of climate; mild, moderate, severe and very severe. These have implications for coating selection.

Very severe exposure conditions are experienced in:

- coastal areas or industrial regions with significant atmospheric pollution and where complete immersion in water or chemicals may occur.

Paint systems which are required to withstand severe exposure have to possess the highest durability. Typically, this necessitates the use of paints based on chemical-resistant binders such as the epoxide, polyurethane, vinyl or chlorinated rubber resins, and these materials are used in situations where limited access makes maintenance difficult, or when cost considerations dictate a long maintenance-free service life.

Severe exposure conditions are experienced in:

- coastal areas subject to salt spray (up to 3 km inland), and non-industrial and with average rainfall;
- inland industrial areas with significant atmospheric pollution;
- areas with high levels of rainfall.

Conventional coatings can be used, although they will have a considerably shorter service life than in moderate conditions. Where a longer service life is required, specialist coatings may be necessary.

Moderate exposure conditions are experienced in:

- semi-coastal areas (3–10 km inland), and non-industrial areas with average rainfall;
- inland areas (more than 10 km from the coast), and urban or light industrial areas with mild atmospheric pollution, but not in close proximity to the source of pollution.

Mild conditions are found in:

- inland areas (more than 10 km from the coast), and non-industrial areas with average rainfall;
- interiors subject to condensation, such as kitchens and bathrooms, or interiors where there is a source of pollution.

Paint systems for moderate and mild exposure regions are typically based on alkyds, oleoresinous varnishes and emulsion resin binders. However, it is not uncommon for a paint system to be required to exhibit a greater durability than is suggested by its immediate environment. Typically, interior paints that need to withstand abrasion or frequent washing fall into this category, and here high-durability paints must be specified.

The ability of paint films to withstand location effects is primarily dependent upon their chemical resistance, a characteristic dictated by the nature of the ingredients used in the formulation.

Most types of coating systems are resistant to the dilute acidic environments which can occur in certain exposure conditions, e.g. in industrial conditions or an urban conurbation where sulphur-containing fossil fuels are burned. However, the presence of sulphur oxides in the atmosphere can cause increased drying times and premature loss of gloss in recently applied air-drying, oxidative systems, although in general these effects do not reduce the effectiveness of coating performance.

The formulation of paints, in particular those based on oleoresinous varnishes and alkyds, are susceptible to attack in dilute alkaline conditions. Where alkali resistance is required, alternative resin types such as the polyurethane, chlorinated rubber and vinyl copolymers have to be used. In situations where air-drying oxidative formulations are required, alkyds modified with these more resistant resins can be used to produce resistant paint systems. However, resistance to concentrated alkalis is more difficult to achieve, since this medium tends to attack all coating types, particularly on prolonged immersion.

Coatings resistant to oils and organic solvents are also required in certain service situations; as a generalisation, coatings with a complex cross-linked structure provide the highest resistance. Paint films can suffer severe attack from solvents, especially chlorinated hydrocarbons (that are the basis of many paint strippers), ketones, esters and aromatic hydrocarbons. Non-convertible coatings, such as chlorinated rubbers, rarely provide good resistance to solvents.

Pollutants, in particular, acids and alkalis, can also adversely affect the pigmentation of paints. Oils and solvents generally have little chemical effect on pigments, although with certain organic pigments the phenomenon known as 'bleeding' can occur on overcoating. This effect is caused by solvents in the freshly applied coating solubilising the pigmentation in the underlying film. This solubilised pigment can be carried through the freshly applied film as it dries, usually resulting in a patchy surface discoloration. The phenomenon of 'bleeding' is especially prevalent where aromatic hydrocarbon, ester and ketone types of solvent are present in the paint formulation.

Influence of weather on coating durability

The effect of weather on paint films is often marked, and can lead to a rapid degradation of coating systems. All of the components of weather—insolation (solar radiation), moisture and temperature—can influence paint film performance. Furthermore, since weather is a complex combination of these components, interactions can occur that render the durability implications of particular combinations of these influences severe, whereas others are relatively mild.

Resistance to ultraviolet radiation.   Prolonged exposure to sunlight results in the rapid degradation of many types of paint films, and this is primarily attributable to radiation in the ultraviolet wavelengths. Ultraviolet degradation of paint films is a complex process involving both an increase in the cross-link density of the internal molecular structure of the film binder and, concomitantly, a tendency for certain constituent structural bonds to rupture. The result of this process is that the film becomes tough during the early stages of exposure. Eventually, however, the film becomes brittle and cracks, and ultimately, as water permeates through the film, it loses adhesion and delaminates (flakes) from the substrate.

To offset this process, it is common practice to add pigmenting materials with the ability to absorb ultraviolet radiation to paints designed for exterior exposure. Titanium dioxide is an example of such a pigment, and it functions by both absorbing the ultraviolet light and reflecting it away from the film, thus protecting the polymer. Invariably with this pigment, backscatter from pigment particles in lower levels of the film results in a certain amount of polymer degradation, and this process releases small amounts of the pigment from the matrix. This reaction, known as 'chalking', is exhibited to varying degrees by all types of titanium dioxide, but it can be used to provide a degree of self-cleansing within coatings. It is, of course, detrimental to the performance of coloured paints containing titanium, since the colour of the film could alter markedly with prolonged exposure. However, by careful selection of pigments, this chalking process can be reduced to a minimum so that, for example, colour changes attributable to this mechanism will not be detectable for a period of several years.

Selection of the binder type is also important in determining the susceptibility of the paint film to ultraviolet light degradation. Polymers containing benzene ring structures are particularly susceptible to ultraviolet degradation since they strongly absorb in the ultraviolet wavelength region of sunlight, i.e. 290–350 mm. Binders without these structures are inherently more durable, although in many instances other performance aspects may render them unsuitable for use in exterior situations.

Resistance to moisture.   Moisture also adversely affects paint film durability, particularly when it is present in the form of rain or condensation. With porous substrates such as concrete, surface moisture, unless present in excess, is generally not detrimental to durability.

Condensation of atmospheric water vapour or rainfall onto the surface of freshly applied paints can effect a premature loss of gloss in all types of coating system due to disruption of the surface. Normally however, this would not result in any long-term reduction in paint durability. Painting in conditions of high atmospheric humidity, that is greater than 90% relative humidity, generally increases the drying times of air-drying oxidative paint systems. However, the drying process is not entirely suppressed, and provided that the film is not damaged mechanically or by prolonged precipitation of rain, snow, frost, etc. whilst in the wet condition, subsequent durability is not likely to be affected.

Temperature effects.   The combination of low temperature and high humidity can also present drying problems, especially with the aqueous emulsion paint systems where film formation is due to evaporation of the aqueous phase and coalescence of the resin particles. With many emulsion systems, optimum coalescence will not occur at temperatures below about 3°C, and since the film formed is not in a coherent state and will not subsequently become so, the durability will be abnormally low.

Low ambient temperatures can also adversely affect the curing rate of many types of two-pack epoxy- and polyurethane-based paint systems, and certain of these coating types will not form films at temperatures below 10°C without the use of additional catalysts.

Non-convertible coatings, that is those that dry by solvent evaporation such as those based on chlorinated rubber and vinyl resins, will dry at very low temperatures. Accordingly, they are ideally suited for winter use. However, the drying times of these types of coating will be increased at any temperature in stagnant air conditions, since under these circumstances, the solvent released by the coating will tend to blanket the surface and impede the evaporation processes.

Tropical and high-heat/high-humidity conditions can also adversely affect the durability of paint coatings. Exposure to high air temperatures, that is above about 50°C, can result in a rapid embrittlement of paint films. In convertible systems, this would be due to acceleration in the cross-linking rate, whereas with non-converting coatings, this could be due to a loss of plasticiser or residual solvent from the film. Furthermore, the expansion and contraction resulting from the temperature cycling of a substrate can induce cracking within brittle paint systems.

High atmospheric temperature effects can be reduced by the use of white pigmentation. Typically, the surface film temperatures of paints with a white pigmentation are half those with a black pigmentation under conditions of summer exposure.

Performance specification for coatings

It is evident that European Standards, once published, will have a profound impact on the coatings industry and upon those involved in their specification. There will be standard requirements for surface treatments applied to concrete. For the first time, coatings used by public bodies will be able to be specified by CEN standards, and CE marking will be used by manufacturers.

The CE mark will demonstrate that a coating conforms to European Community legislation in respect of compliance with the essential requirements of the Construction Products Directive. Whilst this will be desirable, it should be noted that the presence of the CE mark only indicates that the product meets the minimum legal requirements to be placed on the market; it is *not* a quality mark. Undoubtedly many manufacturers will undertake product certification to provide extra assurance to users about the quality and performance of their product. It is a requirement that any such labelling must be clearly distinguished from the CE mark.

Specifiers must be alert to the new European standards and the requirements that are expected to be in place within the next 3 years, but in the interim, there is considerable best-practice guidance within existing documentation such as BS 6150 (1991), the *Code of Practice for Painting of Buildings*. The general principles for the specification of coatings for concrete are summarised below.

General principles of a coating specification

The general principles of best practice in coating specifications for concrete coatings can be summarised as:

- the substrate must be suitably sound and properly prepared;
- an appropriate coating system must be specified;
- the coating must be applied following manufacturer's recommendations;
- the coating must be applied under suitable weather conditions;
- the coating must be maintained at regular intervals.

This latter point is of great importance, as a decision to paint previously unpainted concrete will introduce a commitment to regular repainting, if only for aesthetic reasons (Fig. 6.5).

These general principles of specification will vary in detail depending on whether the surfaces to be painted are fresh or aged/weathered.

New surfaces

The performance of coatings applied to new cement-containing substrates such as concrete can be influenced by the type of mix, the type of admixture, the time of cure before painting, ambient conditions and the type of surface preparation.

The essential requirements of the Construction Products Directive (89/106/EEC, amended by the CE Marking Directive 93/68/ECC 1/1/95) relate to the performance of 'works', i.e. building and civil engineering works. These were set out in Annex 1 of the Directive. Harmonised standards must be able to demonstrate compliance with at least one of the essential requirements. These are:

- mechanical resistance and stability;
- safety in the case of fire;
- hygiene, health and the environment;
- safety in use;
- protection against noise;
- energy, economy and heat retention.

Fig. 6.5 The former Department of National Health building Alexander Fleming House, London (designed by Ernö Goldfinger 1959–1966), after repair and re-use as an apartment block. The repairs included the application of an elastomeric coating over the previously unpainted concrete. The decision to paint the building was the result of a combination of factors: the large number of patch repairs required, the extent of the micro-cracking and an argument that painting would improve the building's appearance (photo Susan Macdonald).

Depending on the intended use of the product and the particular regulatory requirements, all or some of these may be relevant. The connection between these and the specific products will be made through the harmonised European standards and ETA approvals.

Where building regulations require a specific fire performance for walls, coatings should be chosen from those certified to have the necessary surface spread of flame classification; manufacturers will supply copies of their certificates.

Assuming that the quality of the concrete is good, and that it is not excessively porous, then the most important factor is the surface preparation carried out before coating. All dirt, grease and other contaminants must be removed, and there should be no surface laitance. Surface defects such as blowholes should be made good by filling with purpose-designed filler. The most important factor determining coating selection then is the service conditions to which the coating will be exposed (see below).

Other factors will all influence coating selection and performance to some extent, as summarised below:

- the type of concrete mix is likely to be of little importance;
- the type of admixture may have an effect, especially waterproofing additives, which might adversely affect the adhesion of coatings;
- as with any surface to be coated, the ambient conditions may influence coating selection, and will certainly affect coating application. For example, water-borne coatings are more problematic in adverse weather conditions than equivalent but solvent-borne variants;
- concrete can be successfully coated once it has reached a minimum of 80% of the design strength;
- saponifiable coatings, such as those based on oil-containing binder systems, should not be used on new cement-based substrates unless the alkalinity is low and the substrate is unlikely to become wetted in service;
- coatings of low permeability should not be used on new substrates of high moisture content because of the risk of blistering, and manufacturer's recommendations must be followed.

Aged/weathered surfaces

The most important consideration when painting aged, and especially weathered, surfaces is the condition of the surface, as summarised below:

- surfaces must be thoroughly cleaned to be free of all loose material;
- moulds, lichens or algal growths must be removed, and the affected areas treated with an approved biocide;
- graffiti or other contaminants, which may affect applied coatings, must be removed or effectively sealed in, and specialist advice may be required;
- defects such as hollow rendering and large cracks should be made good using an appropriate filler or mortar mix;
- efflorescence should be brushed off and the source of dampness identified and rectified, since no coating will resist efflorescence;

- coatings should be specified in accordance with the substrate, especially its moisture content, alkalinity and general condition, as noted above, and manufacturer's recommendations must be followed.

Coating specifications

A key factor in deciding on a coating specification is the service environment to which the coating will be exposed. The four climatic categories identified in BS 6150 (1991), as discussed previously, can be experienced under both interior and exterior exposure conditions, but it is assumed here that the primary interest is the exterior situation.

As outlined in the previous sections, the development of coatings specifications is dependent on a number of important factors. This complexity makes establishing generic specifications that are applicable to all circumstances of use extremely difficult. General guidance is possible, however, and examples of how coating selection varies in a systematic manner depending on the substrate condition (in respect of moisture content) are given in Table 6.1.

Table 6.1 was developed for moderate (as defined previously) exposure conditions, where the surface of the substrate is sound. In general, Table 6.1 demonstrates that as the moisture regime in the substrate increases, then the options for coating selection reduce, as indeed does the expected time to first maintenance. The table also clearly illustrates the benefits of ensuring that the substrate is dry at the time of coating, since this condition provides the widest range of coating options, some of which have the potential for providing maintenance intervals of 10 years or more.

Other maintenance treatments for concrete

The following treatments for concrete focus mainly on methods for minimising the corrosion of embedded reinforcement.

Applying inhibitors to concrete

The application of inhibitors is designated as Method 11.3 in DD ENV 1504 (1997, Part 9). The role of corrosion inhibitors in concrete repair is discussed in Chapter 5. These are available in the form of liquids, gels or powders that are used to control the rate of corrosion of the steel reinforcement. They fall into three categories:

Table 6.1 Coating systems for cement-based substrates.

Substrate condition	Coating system	Typical life to first maintenance
Dry	Alkyd gloss and modified alkyd gloss	3–5 years or more
	Emulsion paint, general purpose if suitable for external use	Up to 5 years
	Masonry paints, solvent-thinned (smooth or textured types)	5–10 years or more
	Masonry paints, emulsion-based (smooth or textured types)	5–10 years or more
	Masonry paint, mineral type	Indefinite
Drying	Emulsion paint, general purpose	Potentially as for 'dry' substrates but some risk of earlier failure at higher moisture levels
Some damp patches may be visible	Masonry paints, emulsion-based	
	Masonry paint, mineral type	Indefinite
	Possibly solvent-thinned masonry paints	Potentially as for 'dry' substrates, but some risk of earlier failure at higher moisture levels
Damp	Masonry paint, mineral type	Indefinite
Obvious damp patches	Cement paint	As for 'dry' substrates
	Possibly emulsion-based masonry paints	Potentially as for 'dry' substrates, but high risk of earlier failure
Wet	Cement paint	As for 'dry' substrates, but some risk of earlier failure
Moisture visible on surface		

- anodic inhibitors, which control the anodic reaction in the corrosion cell;
- cathodic inhibitors, which control the cathodic reaction;
- ambiodic inhibitors, which control both the anodic and cathodic reactions.

Anodic or ambiodic inhibitors are the most common. They can either be cast in the concrete at the mixing stage, or be applied on the surface of existing structures.

There are many examples where inhibitors are usefully employed to control the corrosion of steel components in other applications. These include tanks containing corrosive solutions and boilers. In such cases, it is a relatively easy task to monitor and maintain the required concentration of the inhibitor in the liquid. Such a level of control in cases where inhibitors are cast into reinforced concrete is extremely difficult, so the optimum concentration around the steel cannot be guaranteed. An inadequate level of anodic inhibitor could lead to cases of localised underprotection, with the risk of concentrating corrosion in such regions.

Inhibitors applied to the surface of the concrete rely on transportation mechanisms through the pore structure of the cement matrix to reach the steel bars. The degree of penetration of the inhibitor to the required level in the concrete will depend on a number of factors, including the quality of the cover concrete and its dryness, and on the depth of the reinforcement. The degree of penetration to the steel is likely to be variable. Evidence supporting the use of inhibitors in such applications is at present incomplete.

Cathodic protection/cathodic prevention

These permanent techniques are discussed in detail in Chapter 5. These techniques are designated as Methods 10.1 and 10.2 in DD ENV 1504 (1997, Part 9). A European Standard describes some of these systems in detail (BS EN 12696-1, 2000). It further recognises that other new and effective anode systems are likely to be developed in the future for cathodic protection application in atmospherically exposed reinforced concrete. Cathodic protection is normally applied when corrosion of the steel is on-going. Cathodic prevention, however, is a preventative measure, and ensures that the steel is polarised to a level where corrosion cannot be initiated. The process also slows down the ingress of chlorides and replenishes the inhibitive hydroxyl ions around the steel.

Sacrificial cathodic protection systems have been successfully installed in the USA and Latin America on structures submerged in seawater or in splash zones. The method relies on anodes made of metal below steel in the electrochemical series (zinc, aluminium), and fixed or sprayed onto the surface of the reinforced concrete component to sacrificially corrode and protect the steel reinforcement. Trials on atmospherically exposed concrete have been less successful, but the main difficulties of the passivation of zinc anodes in insufficiently alkaline concrete, and the gassing of aluminium anodes, are being overcome making the technique very attractive owing to low installation and maintenance costs.

Cathodic protection/prevention in general is a very versatile technique which can be applied in many different situations. The Highways Agency in the UK has adopted cathodic protection systems, incorporating conductive paint anodes, for the majority of the deteriorated crossbeams of the Midland Links Viaduct sections connecting the M5 and M6 motorways.

Painting reinforcement with coatings containing active pigments or with barrier coatings

These techniques are designated as Methods 11.1 and 11.2 in DD ENV 1504 (1997, Part 9). The active components in pigments could include some form of inhibitor or zinc, but their necessarily low concentrations are unlikely to offer any long-term protection to the reinforcement. It is strongly recommended, therefore, that such coatings applied on the surface of the steel reinforcement should not be relied on as the sole protection against corrosion.

The application of a barrier coating to the steel is not a common repair technique, and is likely to be considered for small specific applications only, or perhaps as a short-term holding/maintenance action taken whilst a full remedial treatment is put in train. It requires exposure of the whole isolated steel bar, thorough cleaning of its surface and the application of a non-conductive and continuous coating ensuring total coverage. The corrosion susceptibility of the steel is high at defects and crevices, which should be avoided at all costs.

When applying any coating on the surface of the steel, consideration should be given to the inevitable restrictions on the choice of future alternative protection techniques such as cathodic protection.

Sheltering concrete components

Because the corrosion of steel in carbonated concrete is controlled by its electrical resistance, methods that can maintain the concrete in a permanently dry condition ensure that the corrosion rate of the steel remains at an acceptably low level. The methods associated with this principle seek to limit the moisture content of the concrete by surface treatments, coatings or coatings or sheltering (DD ENV 1504, 1997, Part 9, Method 8.1.) This principle has primarily been applied to cases where deterioration has been caused by carbonation of the concrete. It may be possible that these methods could in some circumstances reduce corrosion activity in chloride-contaminated structures, but there are limited data to prove this. If such an approach were to be adopted, corrosion monitoring would be essential to establish its effectiveness. The principle can be used either as a remediation option, or as a preventative measure against further ingress of moisture, carbon dioxide and chlorides. Potentially, the method might have value in extending the service life of a structure for other methods of concrete deterioration, such as alkali–aggregate reaction.

Monitoring of the moisture content of the concrete or of its internal relative humidity is strongly recommended. Ideally, this would be supported by monitoring of the corrosion activity of the steel reinforcement.

Each specific application should be considered in detail to ensure that moisture cannot be created by condensation or penetrate the concrete from the ground, faulty drainage or any other source.

This is largely a preventive measure, but in some favourable circumstances it can be usefully employed if the corrosion of the reinforcement is carbonation-induced and the chloride concentration of the concrete is low. Ventilated external cladding can also be considered to maintain the underlying concrete in a sufficiently dry condition in cases where there is little risk of moisture penetrating the hidden concrete component.

Management of maintenance works projects

Modern maintenance management of concrete structures is far more complicated than just repairing and servicing system components. The objectives of maintenance planning must be broadened to encompass the long-term performance aspects of the complete concrete building or structure. In order to be effective, a maintenance plan should exist within, and be part of, a quality operating system as a whole, such as the following International Standard, ISO 9000 series:

Standard Number	Title
ISO 9000 (2000)	Quality Management Systems: Fundamentals and Vocabulary
ISO 9001 (2000)	Quality Management Systems: Requirements
ISO 9004 (2000)	Quality Management Systems: Guidance for Performance Improvement

Within the context of maintenance, failure is defined as the inability of a product, system or component to function in the appropriate manner rather than the inability to function at all. However, this assumes that failure depends entirely on what is defined as the required function, which may in turn depend upon the use, environment or operating conditions of the system. Work carried out before failure is said to be overhaul or preventative maintenance, whilst that carried out after failure is regarded as emergency or recovery maintenance. Some of the considerations encompassed by an effective maintenance plan within a defined quality system such as the International Standard, ISO 9000 (2000) series are listed below.

Control of subcontracted products or services

It is necessary to use accredited products or suppliers in order to ensure their suitability for use, and more importantly in terms of maintenance, to allow the identification and traceability of the materials or services throughout all stages of the maintenance process. Identification and traceability are essential components of a quality system if effective methods of process control are to be applied.

Main contractors may subcontract to smaller local firms to carry out various aspects of the maintenance contract. There can be risks involved in using small- to medium-scale construction businesses. Such companies may be vulnerable to liquidity problems and failure. There may be substantial risks in employing the cheapest, lowest-skilled service providers. Court action against liquidated companies will not complete half-finished work. Accordingly, the project manager needs to consider the skill level and management expertise required for the whole job, and what the implications of possible business failures might be.

Control of metrology, inspection and test equipment

All measuring and test equipment defined for use within the maintenance plan should be controlled, calibrated and maintained. The maintenance plan should refer to the measurements to be made, their accuracy and precision, and the equipment to be used to ensure the necessary capability. Relevant equipment should be calibrated against the appropriate standard(s). Equipment should have documented operating, handling and storage procedures. In order to ensure identification and traceability, it is important to maintain calibration records for all inspection, measuring and test equipment.

Planning for long-term maintenance

Consideration of whole-life costing and life-cycle performance issues reveal that maintenance and repair burdens for a building or structure are effectively defined by decisions made in the planning, design and specification phases of its life-cycle. Accordingly, consideration should be given in the planning, design and specification stages to issues such as:

- correct assessment of the loading regimes and environmental conditions that the concrete will be exposed to;
- making a rational decision on the required design and service lives for the structure;
- correct design and specification of the concrete and the structure including, in particular, its 'buildability';
- good construction practice;
- appropriate in-service inspection, monitoring and maintenance actions;
- correct and early assessment of any deterioration, followed by effective repair or preventative measures.

A number of maintenance options have been discussed that may be appropriate for concrete structures and buildings. A management strategy will be required for each structure. This will need to take account of factors such as:

- the intended use of the structure;
- the required residual service-life of the structure;
- the required performance characteristics, including any change of use or appearance.

The approach adopted should include consideration of the need for long-term maintenance, the requirements for proper inspection of key areas of the structure's components, and the installation of appropriate monitoring procedures. Early warning of impending problems would permit the proper management and timely development of remedial measures that should ensure better control and minimisation of overall expenditure. The selection of the appropriate maintenance strategy will be a response to the requirements of the individual structure.

References

BD 43/90 (1990) *Criteria and Material for the Impregnation of Concrete Highway Structures*. Department of Transport, Highways and Traffic, Departmental Standard, April.

BS 6150 (1991) *Code of Practice for Painting of Buildings*. British Standards Institution, London.

BS8110 (1985) *Structural Use of Concrete. Part 1. Code of Practice for Design and Construction*. British Standards Institution, London.

BS EN 12696-1 (2000) *Cathodic Protection of Steel in Concrete*, British Standards Institution, London.

CEB Bulletin 238 (1997) *New Approach to Durability Design. An Example for Carbonation-Induced Corrosion*. Comité Euro-International du Béton (now subsumed into the Fédération Internationale du Béton). Published by Fédération Internationale du Béton, Case Postale 88, CH-1015, Lausanne.

DD ENV 1504 (1997) *Products and Systems for the Protection and Repair of Concrete Structures. Definitions, Requirements, Quality Control and Evaluation of Conformity. Parts 1–10*. British Standards Institution, London.

DD ENV 1992-1, Eurocode 2 (1992) *Design of Concrete Structures*. British Standards Institution, London.

English Heritage (1999) *Graffiti on Historic Buildings and Monuments. Methods of Removal and Prevention. Technical Advice Note*. English Heritage, London.

HB 10141 (1997) *Buildings: Service Life Planning. Part 1. General Principles*. British Standards Institution, London.

ISO 9000 (2000) *Quality Management Systems: Fundamentals and Vocabulary*. British Standards Institution, London.

ISO 9001 (2000) *Quality Management Systems: Requirements*. British Standards Institution, London.

ISO 9004 (2000). *Quality Management Systems: Guidance for Performance Improvement*. British Standards Institution, London.

Siemes, A.J.M. and Rostam, S. (1997) *Durability, Safety and Serviceability. A Performance-Based Design*. TNO Report No. 96-BT-R0437-001, presented at the IABSE Colloquium on Basis of Design Actions on Structures, Delft, 27–29 March 1996.

Further reading

Ashurst, N. (1994) *Cleaning Historic Buildings. Vol. 2. Cleaning Materials and Processes*. Donhead, Shaftesbury.

Ashurst, N., Butlin, R., Chapman, S., Macdonald, S. and Murray, M. (1994) Research into the suitability of sacrificial graffiti barriers for historic buildings. In: *English Heritage Research Transactions*, Vol. 2. James and James, London.

BS 8221-1 (2000) *Code of Practice for Cleaning and Surface Repair of Buildings. Cleaning of Natural Stones, Brick, Terracotta and Concrete*. British Standards Institution, London.

Mills, E. (1994) *Building Maintenance and Preservation: A Guide to Design and Management*, 2nd edn. Butterworth–Heinemann, Oxford.

Chapter 7

Concrete in the Future

Tony Sheehan & Brian Marsh

Introduction

Concrete in the future will undoubtedly retain many of the characteristics of concrete as we know it today. It will continue to contain cement, aggregate and water, and will still be reinforced to provide tensile reinforcement. Over time, however:

- conventional steel reinforcement may receive increased competition from new materials such as polymer composites and steel or polymer fibres;
- additions and admixtures will play an increasingly important role in tailoring fresh and hardened properties to specific needs;
- detailed mix compositions may change significantly in order to reflect the demands of the construction industry of the twenty-first century.

This chapter anticipates concrete in the future, reflecting on current trends, future design drivers and research that will soon become reality. Concrete in the future will be more durable, will be specified with greater consideration of the environment, and will attain significantly higher strengths. Therefore, concrete in its evolving form will continue to provide ever-increasing challenges and opportunities for repair and refurbishment, and will continue to be a dominant construction material.

Improving current practice—the move to intelligent design and maintenance

Many of the problems associated with the concrete structures of the past could have been anticipated. With this in mind, there is a growing movement to plan for the maintenance of buildings at the time of construction, and to programme actions accordingly. This section looks first at the needs for such an approach, and then at the concepts in practice.

The need for an intelligent approach to concrete maintenance

For concrete structures, there is a need to match:

- the durability to the exposure conditions;
- the expected life to the maintenance regime;
- the cost to expectations.

The motivation to maintain and repair concrete structures will vary by client and by timescale. In general terms, however, it is driven by such concerns as:

- deterioration (leading to a reduction in expected life and potential safety concerns);
- unacceptable appearance (as in the need to up-grade older structures);
- structural safety (when deterioration or increased use lead to overloading);
- environmental impact (where contamination or failure leads to an unacceptable blight on the environment);
- expected life (where a structure may have reached its theoretical life, but where extended use is required);
- health (for example, where the collapse of sewers may lead to either short-term structural concerns or longer-term public health hazards);
- change of use (leading to extended life or different exposure conditions).

Each of these 'drivers' alone can provide a justification for repair and/or refurbishment. The decision whether to refurbish or not often simply comes down to whether the drivers are sufficient to offset the resulting significant costs.

The types of defects prompting repair to concrete buildings vary from structure to structure. As has been discussed in detail in the previous chapters, typical causes of concrete decay include (Institution of Structural Engineers, 1996):

- inadequate concrete cover of the steel reinforcement, leading to carbonation-induced corrosion of the reinforcement;
- inadequate resistance to chloride ion ingress, leading to chloride-induced corrosion of the reinforcement, often exacerbated by lack of concrete cover;
- poor quality construction, leading to cracking, voids, leakage and so on;
- poor detailing or poor maintenance, leading to excessively severe localised exposure conditions, e.g. run-off of salt-laden water.

The technical requirements and design methods to overcome such problems are currently covered by codes of practice, but concrete structures are not always built as designed, nor designed with proper consideration of the latest guidance. At present, clients apply great pressure on the designers to produce the best design at the lowest possible materials cost. Thereafter, the construction process is insufficiently robust to ensure that the as-built structure matches the design, which may itself sometimes be flawed, and numerous defects tend to be built in. The system at present, which has been in place since World War II, has many weaknesses and few strengths. Successful concrete construction will be aided by a more holistic approach whereby the interdependency of design, materials, construction, use and maintenance is taken into account at both the design and construction stages.

Even with the best initial concrete construction practice, new demands can be placed on effectively operating structures. For example, many sewers in the UK were constructed 100–150 years ago. Irrespective of the quality of construction practice at that time, they are now of unknown condition and size, and yet remain vital to the country's health and safety. In 1989, it was estimated that forty billion pounds would be required to replace the UK sewage network completely (Arthur, 1989). In practice, careful monitoring, maintenance and repair, rather than full-scale replacement, has helped to extend the life of these structures.

Building lifetimes may be extended for many different reasons; new technical standards or design codes may leave structures designed to old codes at risk of failing to comply with more recent load-carrying or other requirements (such as resistance to wind or seismic disturbance). Indeed, few current design codes explicitly recognise the need to design for specific lifetimes, despite the fact that most common deterioration mechanisms affecting concrete structures are time-dependent. This will become increasingly important as clients become aware of the need to consider construction costs on a whole-life basis. Therefore, a large number of concrete structures may require up-grading in the near future.

The need for repair and remediation increases for concrete buildings as they reach the end of their technical service life. In some buildings, the primary driver may be the owner's desire to up-grade the structure for aesthetic/market reasons; in others it may be due to failure to design for durability, or the realisation that past codes are now inadequate.

In many cases, it is common for those designing new concrete structures to accept the need for future repair and remediation in order to achieve longevity, and as a result to have lower expectations of design and initial construction. This results, to some extent, in a self-fulfilling prophesy, with inadequate checks to detect defects and weaknesses during construction that later become serious defects.

Maintenance measures for some structures may be limited for different reasons, including access difficulties and the potentially high impact of disruption of the core function of the structure. In these instances, it may therefore be necessary to design for a 'maintenance-free' life; alternative strategies are being developed to meet this objective.

In the UK, the current approach to remediation tends to involve periodic inspection, followed by essential staged repairs or up-grades, justified by cost. Any problems are normally characterised into degrees of severity, and the designer may choose to retain, monitor or repair the structure when it is appropriate and cost-effective. However, the process is random, and is more defined by events such as current market demands or finances than planned in advance.

Rather than waiting for failure before addressing it, the future preferred approach can be divided into:

- existing structures—planned maintenance, involving the consideration of likely defects and programming staged repair;
- new structures—intelligent design, by simply learning the lessons of the past and allowing for them.

Each is considered in turn.

Planned maintenance of existing concrete structures

The Polk County Sciences building was the largest of 12 buildings within Florida Southern College designed by Frank Lloyd Wright (Fig. 7.1). Completed in 1958, it was suffering from deterioration due to waterproofing inadequacies and/or inappropriate maintenance (Fig. 7.2). Here, rather than allowing repair and remediation to be dictated by circumstance, an alternative approach was developed—planned maintenance (Architects' Journal, 1998). The concept is simple enough—recognising and assessing repair needs in order to develop a staged programme of repairs.

Two deterioration mechanisms were particularly prevalent at the Polk Science building:

- corrosion of the reinforcement within the load-bearing concrete blockwork, which led to considerable levels of spalling (Fig. 7.2);
- failure of flat-roof waterproofing (an area of approximately 2500 m²), which led to extensive water penetration.

After a thorough assessment of the future needs, a programme of maintenance was carried out. This facilitated efficient budgeting for the client, and allowed a schedule of necessary repairs to be identified. Compliance

Fig. 7.1   The Polk County Sciences Building at Florida Southern College, designed by Frank Lloyd Wright and completed in 1958 (courtesy Ove Arup Partnership).

with this schedule would help to ensure the durability of the concrete structure.

The planned maintenance programme essentially comprised three stages:

- identification of the repair and remediation work and on-going maintenance required;
- identification of possibilities and constraints on repair and remediation, leading to prioritisation of the work;
- application and field development of the 'data spine management system', a computerised method of storing and retrieving information about the condition of building components.

This exercise pioneered a knowledge management approach, i.e. collecting information and storing it in an accessible manner that will be available to the right people in the right form, in the future (Rajan *et al.*, 1999).

Fig. 7.2   The original precast 'textile' concrete blocks were made with fine aggregate, Portland cement and coquina shell. The blocks are highly porous and permeable owing to the absence of course aggregate and poor curing. They are therefore susceptible to a high moisture content and subsequent deterioration over successive cycles of wetting and drying. Poor cover to the reinforcement (due to inaccurate steel positioning and insufficient grout and mortar covering) has led to spalling of the blocks (courtesy Ove Arup Partnership).

By creating detailed information snapshots of the concrete structure over time, more reasoned and timely repair strategies become possible (Fig. 7.3).

As computers increase their ability to store and access information at speed, increased automation of this technique will become an increasing possibility, allowing preventative rather than reactive maintenance to become a reality.

Intelligent design of new concrete structures

Many building owners and managers currently place unrealistic expectations on their structures, requiring minimal initial material cost, but at the same time a long design life. In the future, a more balanced approach will become necessary; planning for future failure mechanisms and taking action at the design stage to ensure appropriate design life will become standard practice.

Fig. 7.3 Limited replacement of the 'textile' blocks was required as part of the repair programme. Prototype replacement blocks, shown here, included a polymer-modified facing mix with a concrete core (courtesy Ove Arup Partnership).

Major improvements in the quality of construction could be obtained simply by considering the lessons from the past, such as:

- changing practice (e.g. avoiding the use of de-icing salts where durability problems result);
- designing for the inevitable (e.g. installing cathodic protection systems in advance for structures where eventual chloride contamination is inevitable);
- matching the specification (i.e. ensuring the as-built structure matches the design intent);
- intelligent design of details to minimise the severity of exposure and to make building easier.

Intelligent design means considering each structure on the basis of the specific performance requirements and the conditions of use, rather than rigidly adhering to the generalised prescriptive rules in a code of practice. Essentially, intelligent design simply involves avoiding the avoidable and learning the lessons of the past (Friend and Sheehan, 1997). Such procedures should be in place already, but can certainly be implemented in the short term. Intelligent design will consider the degradation factors in the exposure environment and anticipate the deterioration mechanisms. It will consider the required performance of the structure over its required service life. The design, specification and construction are then matched to these requirements.

New design drivers—the challenge to concrete

It is remarkable how little concrete mixes have changed over the past 100 years (see Chapter 2). The structural concrete of tomorrow will have the same basic components as the concrete of today, but the construction industry will constantly demand more of the material supplied to it. The following design drivers are likely increasingly to dominate the future UK construction materials industry:

- safety;
- durability;
- appearance;
- sustainability;
- improved performance;
- value;
- speed of construction.

Balancing these factors will be vital to the future of the concrete industry. Therefore, each is considered in turn in the following section.

Safety

There are many health and safety concerns relating to the use of concrete of various types. For example:

- structural collapse (such as that which occurred at Ronan Point and the Camden School for Girls);
- maintenance and controlled disposal of asbestos cement;
- health problems suffered by operatives through the inhalation of cement dust, caustic burns and the damaging effects of noise and vibration during compaction.

As with any material, the placing of concrete will create risks that must be assessed and managed.

The need to avoid foreseeable health and safety risks is an increasing priority for all materials in the UK in light of the *Construction (Design and Management) Regulations* (HMSO, 1994a). These demand that account is taken of predictable health and safety risks to those building, maintaining, repairing, cleaning or demolishing a structure.

The high profile given to scares in the construction press inevitably raises a number of questions. How safe are the concrete structures we now build? What legacies are we creating with concrete for the future? How predictable are the properties of our concretes in the long term? What risks are present with current concreting materials, and how can they be assessed and managed?

The future specifier will require greater assurances that the concrete supplied to site is not posing a future risk. There will understandably be a continued reluctance to use new materials in structural elements without a caseful assessment of long-term properties. The concrete of the future must take account of this in any new materials and methods that are introduced.

Durability

Structural life expectancies of 100 years or more are increasingly specified in the UK concrete industry. For some high-profile projects, such as the Oxford and Cambridge colleges, lifetimes of up to 500 years have been required by owners of new concrete structures.

Whilst the true meaning of such specifications is often unclear in practice, in terms of the required level of performance over the whole design life, it is clear that durability failures in today's concretes are all too commonplace (Fig. 7.4). High-profile failures of motorway bridge decks due to chloride ingress, foundation failure due to thaumasite attack, and the deterioration of facades due to low concrete cover to reinforcement each lead to the need for significant repair and remediation programmes. Alkali–aggregate reaction is also frequently quoted as a source of deterioration in the construction press. The publicity associated with many of these failures often tends to suggest that concrete is a low-durability material. Perhaps the reality is that although these reports tend to over-emphasise the extent of the problem, the maintenance requirements have not been properly fulfilled.

The key to improving the durability of concrete is a careful balance between material properties, exposure conditions and maintenance regimes. Many building owners and managers, for example, expect reinforced concrete by a seafront to last 120 years without maintenance. Guidance is required to create expectations that are more realistic, and to ensure that owners and managers recognise the maintenance required to ensure long-term durability. Alternatively, developers must be prepared to pay the higher initial costs necessary if the structure is to meet such expectations.

Fig. 7.4 Christ the King Cathedral, Liverpool, designed by Sir Frederick Gibberd and completed in 1967, was built to last 500 years. A major programme of repair in the 1990s addressed a series of problems, including those relating to the mosaic cladding on the concrete ribs (photo Susan Macdonald).

Appearance

Whilst strength and durability tend to dominate concrete specification at present, the aesthetics are of increasing interest to architects. In recent years, architects have rediscovered the flexibility and variety of form, colour and texture that concrete can provide. The Bilbao Metro (precast concrete), the Judges Institute, Cambridge (pigmented coloured concrete, Fig. 7.5), and the Jubilee Line Extension, London (*in situ* concrete, Fig. 7.6), are recent examples that demonstrate the aesthetic versatility of concrete.

Fig. 7.5 The Judges Institute, Cambridge, designed by John Outram, 1995, utilised pigmented, coloured concrete to create a highly colourful and decorative building (photo Susan Macdonald).

The design, production and achievement of visual concrete requires a careful blend of the skills of the architect, engineer, concrete supplier and contractor to provide a concrete with the appropriate blend of appearance, placeability and structural performance. This is a considerable challenge, requiring a careful balance of appropriate aggregate selection, and careful cement selection, and selective use of pigments, release agents and surface finishing techniques (whether tooling, sealing or coating) (Monks, 1988).

There are also varieties of finishing techniques (from acid etching to point tooling) that can be applied to the exposed concrete surface. The application of pigments is increasing in the UK, with several companies offering a range of ready-mixed coloured concretes. For large-scale projects, full-scale test panels are increasingly used to provide a realistic trial prior to visual concrete production on site.

The use of concrete as an aesthetic feature must also provide durable ways to retain its long-term appearance. Care must be taken in the use of visual concrete to avoid perpetuation of previous perceptions of a 'concrete jungle' due to the unscrupulous use of acres of stark, plain concrete surfaces. In some cases, precasting may be the only way to make sure of

Fig. 7.6   Canary Wharf tube station, one of the new stations designed as part of the Jubilee Line extension, designed by Foster Associates in 1999. The cut and cover technique was used to construct the station, with durability, fire resistance and aesthetics as essential design criteria. Columns and the roof were cast on site, with precast concrete paviours providing the floor finish (courtesy Ove Arup Partnership).

important factors such as placing, compacting, continuity of supply, curing and formwork quality, fixing and stripping. It is unlikely that a concrete mix can be developed that will make *in situ* concreting immune to these influences in the short term.

Sustainability

Increasing concerns about sustainable development and the Government initiatives that will continue to be developed in response to this will undoubtedly change the face of the future concrete industry. There will inevitably be pressure on building owners, wherever possible, to (HMSO, 1994b):

1.   re-use existing buildings wherever possible;
2.   design new buildings for extended lifetimes, allowing for change of use within that lifetime;

3. maximise the use of recycled materials;
4. minimise the energy use within buildings;
5. minimise waste during construction and make efficient use of all materials;
6. minimise the use of energy during the manufacture of building components, and during construction generally.

Environmental constraints will result in increased pressure on specifiers to search for extended durability and reduce overall life-cycle costs. More sustainable structures will last longer, require less maintenance and have greater potential for re-use.

It has been estimated that in 1998, concrete and masonry accounted for some 90% of the 20 million tonnes of construction wastes disposed of annually at UK landfill sites (Thomas, 1998). This is approximately equivalent to 10% of the UK's primary aggregate use, and offers a great opportunity for change in the area of recycled and re-used materials. In Europe, draft codes of practice for the use of recycled aggregate in concrete have already been produced, and it is recognised as a high priority in the UK (Dhir *et al.*, 1998).

The introduction of a landfill tax has increased the pressure for change, and whilst a significant proportion of crushed demolition debris can currently be used only for fill, the possibilities for recycled aggregate in concrete is the subject of considerable research. Financial pressures will continue to be a major driver of change in the future construction industry. In particular, reduction of waste will be key part of future UK construction. The need to reduce waste in the industry, embrace lean principles and provide enhanced value for clients has already been highlighted (DOETR, 1998).

Improved performance

In terms of improved performance within the concrete industry, there is a constant debate between the relative merits of *in situ* concrete and the benefits of standardisation and preassembly that may result from the effective use of precast concrete (CIB, 1998). On a broader scale, we persist in the production of one-off *in situ* buildings that lead to on-going concerns in terms of 'as built' properties, consistency and current condition, when precast components may offer increased performance and consistency. However, this must be carefully balanced against the need for architectural diversity and expression.

Such issues have been at the heart of recent initiatives that seek to establish:

- annual reductions in construction cost and time;
- reductions in defects;
- sustained improvements in quality and efficiency.

For the concrete construction industry to remain competitive over the next few years, it will either have to match these challenges at least as well as its competitors, or restrict itself to those parts of a structure it does best, such as foundations, floors, frames and precast facades.

Value

Building owners have the right to expect value for money in the procurement and use of concrete structures. Concrete has traditionally been marketed on the basis of long life and low maintenance, and in many cases it delivers these. There are, however, examples of unfavourable press reports and some bad experiences, particularly with bridges and marine structures, that need to be redressed. Owners, designers and suppliers must consider value over the whole life of the structure and select materials accordingly. For this consideration to be valid, however, concrete must deliver its promises.

Speed of construction

For many building owners, time spent in construction is lost revenue-earning opportunity. The concrete construction industry must rise to the challenge of greatly reduced construction times. This can be addressed to a significant extent by existing measures such as high-early-strength concretes and fast-track management concepts. However, there is a need for improvement in areas such as drying of floors and screeds before they can accept coverings.

Improvements may be possible through the increased use of precast elements and modularisation, whereby much of the fit-out can be performed prior to arrival on site. The future is likely to see increased attention to such improvements.

New concrete materials

Whilst the late 1980s saw the advent of high-strength concrete mixes, it is now widely recognised that future concrete mixes should concentrate on improving overall performance as much as strength. High-performance concretes provide more than simply strength, but they must also pay close attention to such aspects as ease of placement, environmental impact, durability and so on.

Essentially, high-performance concretes contain the same components as conventional concretes, and would be produced by essentially the same means (Concrete Society, 1998) . However, careful selection and proportioning of aggregate and cement, and aggregate and admixtures allow the production of concretes with:

- high strength ($100\,\text{N}/\text{mm}^2$ achievable on site);
- high durability (120+ years design life);
- ease of placement (high workability or self-compacting).

Performance can be further improved by additional or alternative measures such as:

- corrosion-resistant reinforcement (e.g. stainless steel or polymer composite reinforcement, both as conventional reinforcement and as prestressing);
- the selection of cement, and additional cementing materials such as microsilica and metakaolin (to increase chemical resistance);
- additives to inhibit corrosion, to remove the vagaries of curing or to control shrinkage.

Ultra high-performance concretes take high performance concrete to an extreme, requiring the modification of design, production and placing techniques. If applied successfully, however, the potential is enormous and the possibilities endless.

A number of high-strength materials have been marketed under this category; RPC or BPR (reactive powder concrete/béton de poudres réactives) in France, CRC (compact reinforced concrete) in Denmark, and SIFCON (slurry infiltrated fibre) in the UK. Essentially, ultra high-performance concretes combine high fibre contents, with or without conventional reinforcement or prestressing, with very high strengths.

There is little doubt that new trends will continue to be developed in the concrete industry, but the take-up of new ideas will continue to be dictated by any cost and/or technical benefits that result. The traditional conservatism of many engineers is likely to continue to slow implementation. Whether any of the above 'new' materials are 'too good to be true' remains to be seen, but future applications for them will certainly be interesting.

Intelligent/smart concrete structures

Rather than designing concrete structures on the basis that their in-service condition will be monitored by periodic inspection, it is becoming possible to develop an advanced assessment approach by providing continuous *real time* feedback on condition by:

- either building in 'intelligence' at an early stage (e.g. at construction to check that the as-built structure matches the design intent);
- or in the longer term creating 'smarter' structures (involving the ability to monitor, self-diagnose or even self-repair).

The use of optical-fibre-based sensors in the UK to detect grout within prestressing ducts is an excellent example of how 'smart materials' can ensure that defects are avoided during construction (Dry, 1995). Voids in grouted prestressing ducts in post-tensioned concrete bridges lead to serious corrosion risks in the long term, and a method of checking safety during construction is increasingly required by clients. The use of optical fibre sensing technology (Michie *et al.*, 1994) ensures that defects are both identified and remedied at the earliest possible stage before any durability, or indeed safety, concerns arise.

Optical-fibre-based strain sensors have also been used in the UK for continuous *real time* strain and condition monitoring. Such devices will be almost essential for newer materials, such as polymer composites, where there are real concerns regarding, for example, long-term creep performance, requiring them to be monitored to ensure continued structural safety. The use of new strengthening techniques such as composite plate bonding (used in the UK to up-grade steel, concrete, timber and masonry structures, see Chapter 5) is growing. The addition of sensors to these plates will provide the confidence required to ensure safe use and to overcome concerns regarding long-term properties.

TRIP (transformer-induced plasticity) steels, which undergo a progressive, irreversible phase change when strained from a non-magnetic to a ferromagnetic phase, can also be used to impart similar intelligence to concrete structures. By direct measurement of the magnetic state, it is possible to assess the peak strains experienced by the material (and any structure to which it is attached) over a period of time. Hence, by attaching a network of such sensors to a bridge (as is the case on several bridges in Georgia, USA), it is possible to detect peak strains experienced in extreme loading conditions (such as wind/earthquake), and to make informed decisions on its future life (Westermo and Thompson, 1995).

It is also possible to produce networks of such sensory materials in order to take the first steps towards a *smarter* structure. The Winooski 1 dam in Vermont, USA, for example, incorporates a data-monitoring system which reviews a network of embedded sensors and provides real-time feedback on the state of the structure (Huston *et al.*, 1995). However, whilst it is possible to remotely acquire both raw sensor data and video/audio images of the structure in real time, there is some concern that in such smart structures the volume of data created may be difficult, if not impossible, to manage and effectively assess. Conventionally, frequent on-site inspections and maintenance checks would have to be carried out on such structures, often incurring considerable travel and attendance costs. The technology applied at Winooski has enabled researchers in Ireland, Australia, Venezuela, Maryland and Vermont to view the structure simultaneously (via the Internet) whilst video-conferencing with each other. Similar technology has been applied to road bridges in Scotland, offering the potential for real-time information/retrieval of structural performance.

Monitoring structures for the ingress of carbonation, chlorides and moisture, and the detection of corrosion should all be practically achievable within a few years. However, such monitoring is likely to be costly, and will probably be justifiable in only a limited number of high-importance structures. Design against degradation should be sufficient in most 'normal' structures, particularly with the increasingly sophisticated design methods that are likely to become available.

If such smart technologies are applied to larger numbers of structures, considerable potential savings in travel and inspection costs could be realised in the relatively short term. Whilst the creation of a network of sensors is the first step towards smart structures, the ideal smart behavior would involve at least some ability to self-diagnose and to act accordingly. Truly smart concretes would be able to sense change (for example the need for repair) and take some compensating action—ideally self-repair.

The first steps towards achieving this will be the development of smarter materials that provide the appropriate ability on a small scale. Hydrogel materials are already widely used in water systems in the UK, and are able to sense the presence of water and react by swelling up (or conversely contracting when the water is removed). Such materials offer great potential as a means of providing smart seals at junctions; swelling up to resist the leakage of water as and when required.

In the not too distant future, newer generations of self-healing concretes could be developed which would be capable of both sensing damage and automatically effecting self-repair. Researchers at the University of Illinois (Dry, 1995) have already developed a range of hollow fibres which can be filled with resin and then cast within other matrices such as concrete or polymers. Significant cracking of the matrix leads to cracking of the fibre, release of the resin, and in essence self-repair is carried out. The use of such built-in self-repair systems within smart structures would allow considerably longer lifetimes to be achieved.

The adoption of intelligent design and smart structural solutions offer perhaps the most potential for carrying out effective and appropriate refurbishment works from the viewpoints of both safety and cost.

Concrete repair in the future

Although it is hoped that the need to repair future structures will be greatly reduced through improvements in design, materials, construction and maintenance, there will be an increased need for the repair of existing structures as they continue to age.

The last 20 years have seen the gradual introduction of a number of new techniques to concrete repair and conservation (see Chapter 5). Cathodic protection of concrete structures, for example, has gradually become accepted as a mainstream repair method. Techniques unique to concrete, such as desalination, have increasingly been specified over the last 5 years, whilst re-alkalisation and migratory corrosion inhibitors have also increased in use. By its very nature, the concrete construction industry remains conservative. The next 20 years will be seen as the trial period for many of these techniques, and no doubt many more will be developed. The longevity of the new repair techniques must be proven if they are to remain at the forefront of the concrete repair industry. Improved durability of repairs should be a principal driver for the development of new and improved materials and techniques.

Other drivers, such as health and safety issues, will also be present. Asbestos, long seen as a wonder material within the construction industry, is likely to receive a widespread ban over the next few years. The proposed UK ban on all asbestos is likely to lay down severe obligations on clients and building owners. Essentially, companies will be expected to assume that asbestos is present unless they can prove otherwise, and they will be expected to keep written records of all asbestos within the building. Asbestos cement tiles were one of the UK's largest markets for white asbestos; appropriate techniques for the assessment and preservation of such materials are urgently required.

Other prominent recent problems, or perceived problems, include:

- calcium chloride accelerators;
- high-alumina-cement concrete;
- alkali–aggregate reactions;
- thaumasite form of sulfate attack.

The wisdom of hindsight might suggest that some of these were foresee-able. Nevertheless, it should be remembered that concrete technology was already well developed at the time of the construction of the structures that suffered these problems, yet they still happened. It should also be borne in mind that the number of actual structures damaged by some of these problems has been very small in relation to the extent of concrete use. The inference that it might be possible to draw from this is that antic-ipation of these types of problem in the future remains unlikely, but also that major problems are perhaps also unlikely.

The greatest source of future problems is likely to continue to be exe-cution, or workmanship. The increasing sophistication of structures and construction techniques could lead to an increase in such problems.

The future will see greater challenges, but through the use of appro-priate technology, the attitude to concrete repair is likely to shift from 'respond and repair', to 'predict and prevent'.

References

Architects' Journal (1998) Florida Southern College, Lakeland. *Architects' Journal*, **207** (2):50.

Arthur, R.A.J. (1989) Whatever happened to the sewer crisis? *Water and Waste Treatment* **32** (9):54–57, 63, 78.

CIB (1998) *Standardisation and Preassembly*. Construction Industry Board, London.

Concrete Society (1998) Design guidance for high-strength concrete. *Concrete Society Technical Report No. 49*, Concrete Society, Slough.

Michie, W.C. *et al.* (1994) Optical fibre grout flow monitor for post-tensioned rein-forced tendon ducts. In: McDonach, A., Gardiner, P.T., McEwen, R.S. and Culshaw, B. (eds) *Proceedings, Second European Conference on Smart Structures and Materials*, Vol. 2361, pp. 186–189, SPIE, Washington, DC.

Dhir, R., Henderson, N.A. and Limbachiya, M. (eds) (1998) *Use of Recycled Concrete Aggregate*. Thomas Telford, London.

DOETR (1998) *Rethinking Construction—Report of the Construction Task Force*. Department of Environment, Transport and the Regions, London.

Dry, C. (1995) Adhesive liquid core optical fibers for crack detection and repairs in polymer and concrete matrices. In: Spillman, W.B. Jr. (ed) *Smart Structures and Materials. Smart Sensing, Processing and Instrumentation*, Vol. 2444, pp. 410–413, SPIE, Washington, DC.

Friend, C. and Sheehan, T. (1997) Refurbishment of UK civil infrastructure: the benefits of an intelligent approach. In: *Proceedings of the Third International Symposium on Intelligent Renewal of Civil Infrastructure Systems*. World Scientific, Capri.

HMSO (1994a) *The Construction (Design and Management) Regulations*. HMSO, London.

HMSO (1994b) *Sustainable Development – The UK Strategy*. HMSO, London.

Huston, P.D., Ambrose, T.P. and Mowat, E.F. (1995) Internet monitoring of an instrumented structure, Fuhr. In: Matthews. L.K. (ed) *Proceedings: Smart Systems for Bridges, Structures and Highways*, Vol. 2446, pp. 301–306, SPIE, Washington, DC.

Institution of Structural Engineers (1996) *Appraisal of Existing Structures*. ISE, London.

Monks, W. (1988) *Visual Concrete—Design and Production*, 2nd edn. BCA/C&CA, Crowthorne.

Rajan, A., Lank, E. and Chapple, K. (1999) *Good Practices in Knowledge Creation and Exchange*. Centre for Research in Employment and Technology in Europe (CREATE), London.

Thomas, D. (1998) Reuse and recycling of building materials. MSt thesis, University of Cambridge.

Westermo, D. and Thompson, L. (1995) Design and evaluation of passive and active structural health monitoring systems for bridges and buildings. In: Matthews, L.K. (ed) *Proceedings, Smart Structures and Materials. Smart Systems for Bridges, Structures and Highways*, Vol. 2446, pp. 37–46, SPIE, Washington, DC.

Chapter 8
Repair, Remediation and Maintenance in Practice— Case Studies

Introduction

A number of case studies have been selected to illustrate the principles of investigation, repair and maintenance as explained in the previous chapters. These examples depict various repair methods, and explain the investigations and subsequent decision-making process that led to the adoption of the particular repair strategy. The case studies are drawn from the UK and further afield, but have a common theme: each project progressed through a process of professional investigation in order to remedy the problems specific to that building.

All the structures dealt with in the case studies have been recognised as being of heritage value and therefore involved the relevant conservation authority. In some cases, the heritage significance of these structures meant that there were additional constraints and opportunities that had to be factored into the repair and maintenance program. The UK examples are drawn from the author's experience when working with English Heritage.

The repair of two nineteenth-century concrete structures in Sydney, Australia

Lance Horlyck & John Lambert of Sydney Water, Australia

Sydney Water is the body responsible for the management of a large number of concrete structures in the Sydney metropolitan area, including 10 major dams, 200 reservoirs, over 700 pumping stations, 500 km of stormwater channels, 400 km of trunk sewers and over 30 treatment plants. Some of these assets date back to the last century and include one of the first uses of reinforced concrete in Australia, the heritage listed Johnstone's Creek sewer aqueduct constructed in 1897. Unlike many other concrete structures, it is rare to demolish and replace sewer and water infrastructure. Instead, there is the tendency for the existing system to be amplified or augmented as demands increase. In the majority of cases, major assets are expected to continue to function for the foreseeable future, even when they are approaching or have exceeded their originally intended design life.

Historically, Sydney Water were committed to the production of high-quality concrete and developed their own set of standards for design and construction. This has resulted in high-quality and relatively durable structures, the majority of which continue to function adequately. This is also a result of a well-monitored maintenance program. The repair and maintenance of structures by Sydney Water utilises assessment procedures based on life-cycle costing.

Appraisal methodology

The initial assessment of concrete structures in Sydney Water's care is usually a visual inspection—either externally and/or internally—that is carried out as part of regular monitoring of assets. If some form of change or worsening of the state of the structure is detected, a condition assessment, structural appraisal and feasibility study are undertaken to assist in future programming of maintenance and repairs. Detailed investigation work, undertaken as part of the condition assessment, will vary to suit the type of structure, age, current state of deterioration, access, availability of drawings and the expected cause of deterioration. Typically, the investigation procedures may include:

- detailed visual inspection (crack mapping, quantifying deterioration);
- geotechnical investigation (soil and foundation conditions and loadings);
- groundwater aggressivity measurements (pH, chlorides and sulphates);
- cover meter survey (reinforcement location and workmanship);
- electropotential mapping (corrosion potential of reinforcement);
- carbonation levels and chloride contents (environmental ageing of concrete);
- concrete core sampling (concrete strengths and densities);
- Schmidt hammer hardness testing (concrete strength);
- crack monitoring (active or static cracks);
- temperature and movement monitoring (effectiveness of expansion joints);
- gas and humidity monitoring (hydrogen sulphide-related attack).

The condition assessment will also include some form of structural analysis/appraisal in order to:

- assess the sufficiency of the original design;
- determine the impact of the current level of deterioration on structural adequacy;
- determine the levels of deterioration necessary for various types of repair strategies;
- predict when deterioration will reach a stage where there is an unacceptable level of risk of collapse.

The object of the condition assessment is to find out what is causing the deterioration, what is its impact on the structure, how quickly the condition is likely to worsen and how long the problem can be left before it becomes critical.

Repair strategies

Where the investigation and appraisal has identified that intervention will be necessary, a detailed feasibility study is performed. All the different repair strategies are considered and evaluated (including those involving operational changes rather than structural rehabilitation) to establish the best form of repair from a combined financial, maintenance and operational perspective. The following case studies demonstrate this approach.

Johnstone's and White's creek aqueducts—Australia's first major reinforced concrete structures

Johnstone's and White's creek aqueducts were constructed in 1897, in the inner Sydney suburb of Annandale (Fig. 8.1). They were the first major concrete structures in Australia to incorporate reinforcement, and as such are recognised as being of heritage value and are protected by legislation. The success of this work pioneered the use of reinforced concrete in Australia.

The two aqueducts are each approximately 300 m long, and are almost identical. Effluent is transported through a conduit up to 11 m above ground level and supported by a series of arches each 25 m long. The aqueducts were constructed using Joseph Monnier's technique of reinforcement, and covered with a final coat of cement render.

Considering their age, the structures have performed very well. It was not until the 1980s that rehabilitation work began to address some of the obvious problems. This work was carried out for both structures, and was directed mainly to rectifying problems in the conduits. Steel ties in the

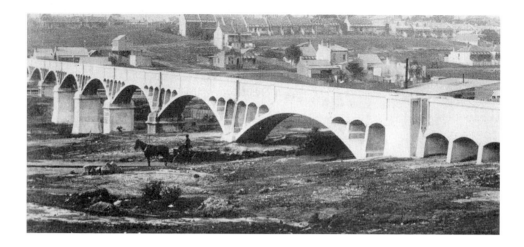

Fig. 8.1   Johnstone's Creek sewer aqueduct, Annandale, shortly after construction in 1897.

conduits' roof had become badly corroded, and were replaced with new stainless steel rods. The conduits were also found to have experienced extensive unsightly leakage, and as a result, liners were fitted internally.

More recently, corrosion extended to the reinforcement in the major arches. The corrosion was so extensive in some arches that the surrounding concrete was spalling and was falling to the ground. The situation was rectified with recently completed repair and remediation.

Investigation

The following investigations were carried out on a representative span of each aqueduct:

- visual survey of the concrete;
- delamination survey;
- covermeter survey;
- depth of carbonation;
- chloride content analysis;
- electrochemical potential survey;
- break out and inspection of reinforcement;
- concrete compressive strengths.

Figure 8.2 describes the survey recording methodology.

These investigations generally gave results that corroborated each other, and in summary indicated the following trends:

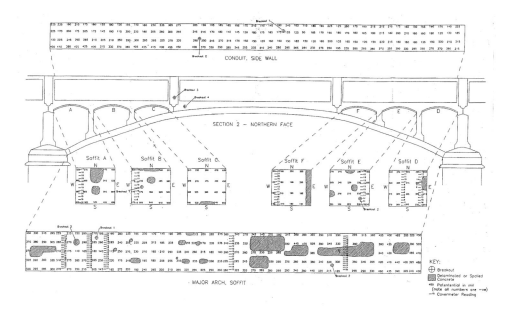

Fig. 8.2   Condition survey of a typical arch at Johnstone's Creek aqueduct, showing cover-meter and electrochemical potential readings, and areas of delaminated or spalled concrete (courtesy Taywood Engineering).

- the depth of carbonation was greater than the average cover to reinforcement;
- electrochemical potential measurements were high, indicating that more than 50% of the soffit reinforcement was actively corroding (which was confirmed when repairs were carried out);
- the chloride content values were relatively high in areas where reinforcement had corroded.

In addition to investigations of the concrete, the other procedures carried out were:

- a heritage assessment;
- a level survey to check for settlement;
- movement monitoring at expansion joints;
- an internal inspection of conduits;
- a foundation inspection;
- a detailed structural analysis.

The investigations and analyses determined several key principles.

- The aqueducts should be restored to their original appearance for reasons of heritage significance.

- The aqueducts were not properly designed to cater for thermal expansion and contraction, as only the walls of the conduit were provided with vertical expansion joints at each pylon. As a result, cracking of the conduit between expansion joints had occurred, causing unsightly external staining due to the efflorescence and sewage leakage.
- Horizontal transverse tie rods connecting the walls of the conduit had corroded so badly in the roof that they had been replaced with stainless steel ones about 10 years earlier. Some of the tie rods in the floor had also corroded, causing cracking and separation of the unreinforced floor.
- The lack of top reinforcement in the major arches, together with relatively weak concrete (12 MPa), meant that the arches did not comply with current design standards. This, combined with the degree of corrosion already present in at least 50% of the arch soffits and cover to reinforcement, being much less than current durability criteria, confirmed the urgent need to carry out repairs to the arches.

Repair procedures

Various options were considered for repairs to the aqueducts, including patch repairs and also the addition of new structural arches under the existing major arches. However, in the end it was decided to carry out full repairs to the aqueducts. The main repair aspects that were initiated are described below.

- For the major arches, the entire arch soffit was rehabilitated to increase cover from 12 mm to 30 mm in order to satisfy the Australian standards. The soffit repair was staged in two equal widths for the full span of each arch to ensure that structural integrity was not jeopardised.
- Removal of the existing render and concrete cover to the arch soffit reinforcement using water jetting was not successful owing to the high noise level and excessive gouging of the concrete. A more conventional concrete removal technique using small jackpicks was adopted to ensure that render and concrete were removed evenly, that the full perimeter of existing reinforcement was exposed, and that a further 15 mm space behind the bars was free for proper penetration of the shotcrete. Significantly large areas of existing soffit reinforcement were found to have corroded beyond an acceptable 20% of the bar's cross-sectional area.
- The concrete in the major arches was reinstated with appropriate cover by a dry shotcreting process that was frequently tested for its integrity by pull-off testing and was found to be very effective.

Fig. 8.3   Johnstone's Creek aqueduct after repair (photo: Susan Macdonald).

- In the pylons, minor arches and external faces of conduits, drummy or badly rendered concrete was removed and corroded reinforcement was grit-blasted or replaced. A bonding mortar was applied to concrete and reinforcement in the repair area, followed by application of a high-build repair mortar (both consisting of a cementitious polymer and silica fume-modified mortar). A cementitious polymer modified seal coating was applied over the entire surface area to provide a uniform appearance (Fig. 8.3).

Centennial Park reservoir—a mass concrete structure

The heritage-listed Centennial Park reservoir was completed in 1898 and utilised the latest available technology at the time of its construction. One of the oldest surviving reservoirs still in use, it stores 80 Ml of water for a large suburban area of Sydney and covers an area of 1.4 ha. In 1992, major repairs were undertaken on the southern perimeter to stabilise the walls and roof. Although the repair work itself raises many interesting issues about the use of modern-day durable concretes for water-retaining structures, of more interest is the composition of the original concrete roof. This is constructed of mass concrete groined arches supported on brick columns on a 6 m × 6 m grid. Wrought iron tie bars are located in

Fig. 8.4 Centennial Park reservoir, showing mass concrete groined arches and brick columns (photo Sydney Water).

orthogonal directions between columns at the springing point of the arches (Fig. 8.4).

An important part of the assessment process for the repair and remediation work was an evaluation of the condition of the existing roof. Core samples extracted from the roof indicated that it comprised lightweight concrete (1450 kg/m^3) of high porosity with a very low compressive strength. The fines content on the concrete was low, and the aggregate was found to contain inclusions such as charcoal. A coal-fired power station used to provide electricity for trams was obviously a source of cheap aggregate during the original construction. Analyses indicated that the double-arch profiles were stable. The only areas where problems were experienced was at the perimeter, where corrosion and subsequent failure of tie rods and truncated panels at the curved ends resulted in movements and tensile forces that caused cracking in a number of panels. Apart from works to address the problems at the edge panels (shown in Fig. 8.5), programmed maintenance procedures including regular visual monitoring was all that was required for the rest of the roof. This project indicates that mass concrete structures, composed of what would now be considered substandard concrete, can produce structures with exceptional life expectancies.

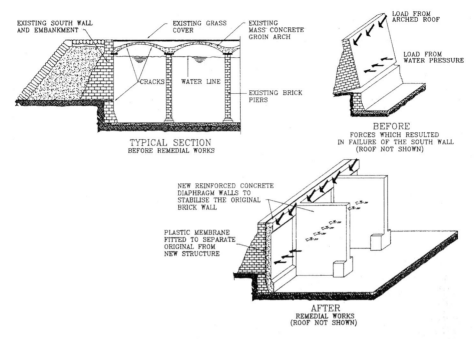

Fig. 8.5 Centennial Park reservoir, showing the strengthening of perimeter walls (courtesy Sydney Water).

Conclusions

These two case studies show the durability of high-quality historic concrete. Both these structures belong to the pioneering era of concrete construction, and have undergone only minimal maintenance until the recent repair and remediation projects.

The repair strategies developed were based on life-cycle cost–benefit studies, and maintenance and operational requirements. By adopting this approach, Sydney Water have been able to extend the operational life of many important concrete assets, thereby avoiding the far more expensive and often unacceptable alternative of replacing the structure.

The repair of Mies van der Rohe's Promontory Apartment Building, Chicago: a multi-phased approach

Paul E. Gaudette, Harry J. Hunderman & Deborah Slaton

Introduction

The Promontory Apartment Building was designed by Ludwig Mies van der Rohe in 1949. Located at 5530 South Shore Drive, Chicago, it was Mies van der Rohe's first large-scale commission outside of the Illinois

Institute of Technology campus and his first constructed high-rise building. The building was also Mies's first collaboration with the developer
Herbert S. Greenwald (1915–1959), who became one of his most important clients. Pace Associates were the associate architects for the project,
Holsman, Holsman, Klekamp & Taylor were the consulting architects,
Frank J. Kornacker was the structural engineer, and the general contractor was Peter Hamlin Construction Company (Architectural Forum,
1950).

The 21-storey Promontory Apartment Building is typical of the post-
war trend toward straightforward design with an emphasis on efficiency,
low cost and a clearly expressed structure. The exposed reinforced concrete frame of the building is infilled with light-coloured brick panels (Fig.
8.6). The concrete frame is emphasised by the projection of the columns
beyond the exposed floor slabs. The windows are aluminium-framed.
This is an early architectural use of aluminium following the popularisa-
tion of this material for wartime aircraft construction.

The original design for the Promontory Apartment Building included
a curtain wall of steel and glass that was the forerunner of Mies's later
buildings at 860 and 880 North Lake Shore Drive, Chicago. However,
the post-war steel shortage dictated the use of concrete.

Fig. 8.6 Front and side elevations of the Promontory Apartment Building,
Chicago (photo Paul Gaudette).

The current exterior conservation works were prompted by water leakage and the deterioration of elements of the facade, including the exposed concrete, masonry walls and joint sealants. The challenge of this project was to execute repair work that would perform well and match the appearance of the existing original materials. This case study summarises the phases of the project, focusing on issues related to concrete, including the investigation, the laboratory analysis of building materials, development of the concrete repair mix, trial repairs and the repair program.

Building description

The structural system for the Promontory Apartment Building consists of a frame of concrete beams and columns. The interior floor slabs are supported by a one-way floor joist system that spans into the floor beams. The perimeter components of this structural concrete frame also act as part of the exterior facade. The columns are buttressed and step back, or reduce in cross-sectional area, at the sixth, eleventh and sixteenth stories of the building. In plan, the perimeter columns extend past the slab edges.

Brick masonry was used as infill between the structural concrete elements of the exterior facade. A typical bay is shown in elevation in Fig. 8.7. The infill extends from a recessed curb along the top of the concrete slab edge to the bottom of the window sill. The head joints located along the bottom course of bricks were left open to act as weep holes for the

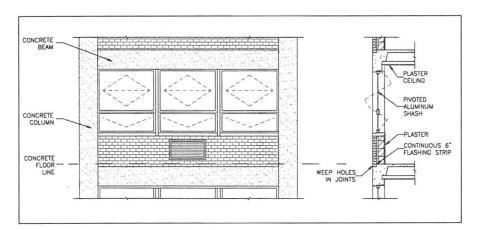

Fig. 8.7 A typical bay. Brick masonry was used as infill between the concrete frame elements. Aluminium pivot, and hopper windows extend from the concrete spandrel beam to the concrete sill above the masonry. This section is excerpted from the original Mies van der Rohe construction drawings.

masonry walls. Copper flashing was installed at the base of the masonry wall to assist in directing water out of the wall through the weep holes. Air conditioners were later installed in the brick masonry spandrels, based on details provided by Mies in 1966.

The aluminium-framed windows have a fixed upper sash that can be unlocked and pivoted for indoor washing, with an inward-opening hopper below. Fresh air is provided through the hopper windows. Steel, aluminium and stainless steel were considered for the window frames. Aluminium, not yet widely exploited in curtain walls but readily available following expansions in the industry during World War II, was finally chosen.

Appraisal and investigation

Although there have been a number of repair programs at the Promontory Apartment Building since it was constructed, deterioration of the concrete, brick spandrels, sealant and windows led to a comprehensive investigation of the building's exterior. The primary types of distress were related to cracking and spalling of the concrete and deterioration of the mortar joints and sealant joints. The purpose of the investigation was to determine the causes of the deterioration and to develop a plan for the conservation of the facades. The investigation consisted of a review of available documentation, an initial visual inspection, a hands-on, close-up examination, and laboratory analyses of building materials.

The appraisal and investigation of the exterior was designed to evaluate the existing conditions and distress. Original drawings and field notes were available, but not the original specifications. After a visual examination of the facades from grade, interior spaces and the roof areas of the building were selected for detailed examination. The detailed inspection was made from a suspended scaffold at four representative bays. During this inspection, areas of delaminated concrete were removed because of safety concerns. Selected areas of brickwork were opened up to determine the existing conditions, the causes of deterioration and the as-built conditions. Samples of the concrete, mortar and sealant were also removed for laboratory analysis.

The building appeared to have undergone several previous phases of repair over its life, but these were not documented in the building records. During the detailed examination, openings were made to assess the condition of these repairs and to determine the repair techniques used.

From this examination, it was determined that crack repairs had consisted of the application of a slurry of epoxy and sand over cracks in the spandrel beams. The previous repair materials appeared brown in colour, and did not match the original concrete in colour or texture. These repairs

Fig. 8.8 Removal of a delaminated concrete repair: this repair was installed to the level of the reinforcing steel only (photo Paul Gaudette).

were very noticeable from street level, had cracked over the original crack, and had a negative impact on the overall appearance of the building (Fig. 8.8).

The previous repairs were found to be cracked, delaminated and debonded from the original concrete. It was obvious that patch material had been applied without removing the original concrete around the reinforcing steel. Typically, the repairs consisted of the installation of a trowel-applied mortar over corroded reinforcing bars. These repairs were also very noticeable from street level.

New sealants had been applied over existing sealants with little, if any, surface preparation. Masonry panels had been repointed by removing loose mortar and applying new mortar over the top surface, filling some of the weep holes at the bottom of masonry panels with sealant and mortar, and installing sealant in the bottom joint of the masonry panels.

Laboratory analysis

After the field investigation, some materials were analysed to determine their components, composition and causes of deterioration. Laboratory studies of the concrete included petrographic evaluation following ASTM C856-95e1 (1995) and tests to determine the air content, water–cement ratio, cement content, general aggregate identification, carbonation depth and chloride content.

The petrographic evaluation was performed to provide a general identification of the components and aggregate of the original concrete. This information was needed to develop a mix design for the repair concrete. The petrographic studies revealed that the original concrete was made with a natural coarse gravel aggregate that is petrographically similar to the 'Elgin gravel' that has been used in the Chicago area for many years. The fine aggregate was composed of siliceous sand.

Concrete deterioration in building facades is generally related to two principal causes: corrosion of embedded steel and chemical degradation of the concrete matrix. At the Promontory Apartment Building, deterioration was primarily caused by corrosion of the inadequately protected embedded steel. In the samples evaluated as part of the investigative work, the carbonation was found to be typical for a building of this age and exposure, and was not considered to be a major factor in deterioration.

Laboratory analyses of the concrete samples also indicated a relatively low cement content (4–4.5 bags of cement per cubic yard of concrete) and a variable, moderate to moderately high, original water–cement ratio. In addition, a microscopic examination revealed that the variation in the amount of exposure of the aggregate particles on the surface was due to the differential dissolution of exposed cement paste. The more that water 'scrubs' the surface of concrete, the more the cement paste is dissolved and washed away, exposing the aggregate. This weathering is typical of concrete surfaces.

In the samples examined, the body of the concrete was found to be sound and intact, with disruption confined to surface erosion and spalling. The spalling observed was associated with the expansive forces created by corrosion of the embedded steel reinforcement, as discussed in Chapter 4. Chloride levels in the samples were found to be at, or slightly above, the threshold at which the corrosion of embedded steel is promoted.

The concrete was found to be air-entrained, with an air content of approximately 3%. This is considered in the low range. The use of air-entrained concrete in a structure of this period is not typical; air-entrainment, which has been found to improve the freeze–thaw durability of concrete, did not gain popularity in the construction of high-rise buildings until much later. Concrete used in the structural frame of the Promontory Apartment Building was specified to be air-entrained to facilitate its placement. This was probably owing to the concern for a consistent appearance of the concrete portion of the exterior facade (Architectural Forum, 1950).

Conservation strategy

The aim of the conservation project was to repair the exterior concrete, address any deterioration of other exterior elements of the facade, and reduce the rate of future deterioration of exterior building materials by reducing the rate of moisture infiltration into the facade. The primary objective of the repairs was to use materials and techniques that would be sympathetic to the existing facade and perform well. Finally, the repair design needed to meet the installation tolerances used in the original construction. The concrete and brick were meticulously installed in extremely straight lines, and with the very low tolerances typical of Mies-designed structures.

In order to achieve these aims, the project was organised in three phases.

1. Development of trial repair materials and procedures.
2. Performance of trial repair work at one drop of the building.
3. Performance of repair work on the rest of the building facade.

Trial mixes and repair techniques were evaluated to determine how best to match the original appearance while providing a durable repair. The implementation of repairs at one trial drop permitted a technical and aesthetic evaluation of the completed repairs, and an assessment of the scope of work and the contractor's procedures. Information gathered in the first two phases was utilised in refining the requirements for the project.

The work completed under each phase consisted of the steps listed below.

Phase 1:

- developing trial repair materials and procedures, which involved determining the repair materials, and developing and testing trial mixes;
- developing repair procedures and repair techniques;
- performing finishing samples, using various techniques;
- selecting repair materials and finishing techniques to match the existing concrete;
- selecting a system to reduce the amount of moisture penetration into the concrete.

Phase 2:

- performing trial repair work at one drop of the building, which included using trial repair materials and techniques;
- evaluating the work, modifying procedures, and repeating trial repair work as needed;
- performing repair work on the trial drop;

- modifying repair materials and techniques to adapt to actual as-built conditions.

Phase 3:

- performing repair work on the remaining portions of the facade, which included incorporating lessons learned in Phase 2;
- performing repair work at a trial repair area;
- performing work on the remaining areas of the building.

The repair program

Phase 1 focused on developing a mix design for a repair concrete to match the original. The first problem was to identify the aggregates, sand and cement used in the original concrete. Laboratory analyses revealed that the aggregate was a natural gravel composed primarily of dolomitic limestone. The natural sand was composed primarily of quartz and chert, with smaller amounts of limestone and minor amounts of shale. The cement was buff/white in colour. Fortunately, the majority of these materials are readily available locally. The buff-coloured cement, not commonly used or produced today, was more difficult to obtain.

Testing during the project helped to maintain consistency in the repair materials. The testing parameters were developed during the trial repair phase, so that they could be evaluated and adjusted prior to the implementation of full-scale repairs. The initial parameters developed for laboratory or field testing included slump air content and compressive strength. Slump (measured by ASTM C143) is a measure of the concrete's workability and consistency, which determines whether it can be consolidated properly within the forms and repair areas. A slump test is performed by placing fresh concrete into a cone, removing the cone and measuring the vertical distance that the concrete settles. Air content is measured by a pressure meter (measured by ASTM C231-97e1, 1997) and indicates the entrained-air content (the incorporation of microscopic air bubbles), which provides protection for the concrete against damage due to cyclic freezing action. The concrete is air-entrained by the addition of an admixture during the mixing process. Compressive-strength testing (by ASTM C39/39M-01, 2001) confirms that the concrete meets the required strength for the installation.

In order to match the existing finish, texture and colour of the original concrete, a conventional air-entrained concrete was selected instead of a polymer-modified repair concrete, which is now typically used for external repairs. Polymer-modified patches are generally acceptable where a coating is to be used to cover the entire repaired surface. However, no coating was previously used or was intended to be used on the facade of the Promontory Apartment Building.

All concrete repair materials were placed into formwork with a minimum depth of approximately 2 inches (50 mm). Trowel-applied thin patches, using mortar repair materials without forms, were not used. Formed patches with greater depth and coarse aggregate provide more room for the proper placement and consolidation of the repair concrete, and usually result in a more consistent, durable repair. In addition, all patches were anchored to the original reinforcing steel by the excavation of existing unsound concrete back to sound concrete a minimum of 0.75 inches (19 mm) beyond the depth of the exposed steel within the patch. This provides a more substantial mechanical attachment to the structure. In the case of misplaced original steel, additional reinforcing steel was added to provide even greater attachment. The placement of the concrete was also improved by using both internal and external vibration techniques. These techniques can be used with formed patches but not with trowel-applied patches.

Approximately twenty 1 foot × 1 foot (300 mm × 300 mm) samples were prepared in forms, separate from the building. The samples incorporated a variety of mixes, with different proportions of buff-coloured cement and aggregate components, and different finishing techniques. It was difficult to match the appearance of the existing concrete because of the varying degrees of paste erosion and resultant aggregate exposure. Several of the trial mix samples are shown in Fig. 8.9. Finishing techniques and procedures were developed to allow the contractor to vary the exposure of

Fig. 8.9 Trial repair mix samples adjacent to the original concrete façade (photo Paul Gaudette).

the aggregate in the concrete to match the appearance of the original facade. Some of the surface finishing techniques evaluated included the application of a surface retarder, sandblasting, water blasting, low-pressure water blasting, stone rubbing and hand brushing. Samples that utilised a surface retarder were rejected because the resulting appearance was too even and exposed too much aggregate compared with the original concrete. The most effective finishing techniques involved a combination of very light water blasting and hand finishing. Once the finishing techniques were refined, the proportions of aggregates and of the buff/white cement mixture were adjusted.

Previous crack repairs using an epoxy material were unsuccessful in bridging cracks and in preventing moisture from entering the concrete. As a result, embedded reinforcing steel in the area of the crack continued to corrode, and deterioration continued in areas of previously repaired cracks. To correct this situation, the cracks were routed to 3/8-inch (9.5mm) thickness and filled with a sealant. The sealant colour was selected to match the adjacent concrete as closely as possible.

To reduce moisture penetration and deterioration further, a penetrating sealer was applied to the concrete facade. During the past few years, new penetrating sealers have become available that make fine cracks and pores in the concrete resistant to water while allowing any water vapour that does enter the concrete to escape. These new penetrating sealers include silanes and siloxanes. A silane is a very small molecule, called a monomer, whose size approaches the size of the pores in many stone and concrete substrates; consequently, the silane is able to penetrate deeply into the pores. The silane reacts chemically with the surfaces of the pores and makes them water-repellent, or hydrophobic. A siloxane is a prepolymer made up of a number of monomer units; it is therefore larger than a silane. The siloxane is also chemically reactive and bonds to the surfaces of the substrate pores. Neither silanes nor siloxanes form a film, and they do not noticeably alter the appearance of the facade. Some silanes or siloxanes are used in combination with a film-forming acrylic topcoat. In addition, some formulated products combine a silane or siloxane with a film-forming polymer. Film-forming products may alter the vapour permeability of the substrate, and may also affect the texture, reflectance and overall appearance of the surface. The penetrating sealer selected for use on the Promontory Apartment Building is a silane-based product without a film-forming topcoat.

After finalising the repair mixes and finishing procedures, a mock-up of the repairs was performed on a portion of the building. The location of the mock-up was selected to be both accessible and unobtrusive. The mock-up was executed on an area of concrete at one floor level across one bay. Refinements to the mix and placement procedures were made during the mock-up. For example, the placement of the mix design selected during the sample preparation was difficult because the concrete did not flow readily, resulting in inadequate consolidation of the concrete within the patch. A small amount of water was added to the mix to facilitate the placement of concrete in the formwork, and other adjustments were made as the mock-up proceeded.

Surface preparation is one of the most important components of any concrete repair. The steps followed at the Promontory Apartment Building are fairly typical of concrete repair work, but were slightly more aggressive than normal in the removal of concrete within the patch area in order to provide better encasement of the reinforcing steel within the repair material, and to improve the performance of the patch (Fig. 8.10). The procedure for surface preparation was as follows:

- remove loose and unsound concrete from the exposed portions of the columns and spandrel beams;
- bevel saw-cut the perimeter edges of the repair area to a depth of 1 inch (25 mm) and approximately 1 inch beyond the visible corrosion of the reinforcing steel;
- chip the concrete within the patch area to a minimum of 0.75 inches (19 mm) deeper than the reinforcing steel;
- sandblast and air-blast the patch area to clean away laitance, dirt and other debris from the exposed concrete;
- sandblast and air-blast exposed reinforcing steel to remove rust, dirt and other debris;
- inspect the exposed reinforcing steel for loss of cross-sectional area due to corrosion and replace as required;
- install supplemental steel, as required;

Fig. 8.10 Completion of the preparation works in the repair area prior to the installation of the formwork (photo Paul Gaudette).

Fig. 8.11 Installation of formwork and a port for ease of concrete placement (photo Paul Gaudette).

- apply a corrosion-inhibiting coating to the exposed reinforcing steel within the patch.

It was extremely important to match the surface profile and finish of the adjacent concrete. Heavy grinding of the patch surface would have prevented a good match with the finish and texture of the adjacent concrete. The procedure for placement and finishing was as follows:

- install formwork at all repair areas to match the existing profile of adjacent concrete (Fig. 8.11);
- test concrete for conformance to specifications;
- place concrete into forms using internal and external vibration techniques;
- after approximately 24h curing, remove the forms;
- expose aggregate at the exterior surface of the new concrete with a combination of hand brushing and low-pressure water blasting to expose the aggregate to a depth that resembles the original concrete adjacent to the repair area;
- cure the repair concrete by installing plastic over the repair area for a minimum of 7 days;

- test concrete for conformance to specifications;
- apply penetrating sealer to both the original and the repair concrete to reduce the amount of water penetration.

Conclusions

During the sample repairs executed on one bay of the facade, procedures and materials were adjusted to achieve a concrete repair that matched the adjacent original concrete in appearance, and met the established criteria for good concrete repair practice. The multi-phased process of investigation, laboratory analysis, trial samples, mock-ups and full-scale repairs allowed refinements of the repair design, the maintenance of installation procedures and the implementation of quality-control measures as the project progressed. A conservation approach was used to guide technical and engineering decisions, resulting in repairs that perform to modern practice standards and are aesthetically successful.

Burleigh J.M.I. School, Hertfordshire—repairing precast panels

Introduction

The now-listed Burleigh J.M.I. School in Cheshunt, Hertfordshire, was designed in 1947 by the Hertfordshire County Council Architect's Department (Fig. 8.12). It was one of the schools built as part of the innovative and extensive school building programme in Hertfordshire immediately after World War II. This programme had a major influence on the design of schools across the country. The architects developed a system of modular, prefabricated components that could be assembled in various ways.

Burleigh was one of the first schools to utilise this precast concrete cladding system. At this time, the technology was moving so quickly (about ten of these schools were built per year from 1948) that advances and improvements to the construction system were implemented on the next building to be constructed, and therefore the details are not uniform from school to school.

Cladding failure

The unpainted concrete panels used to clad the buildings at Burleigh were only 2.5 inches thick (63.5 mm), and were butt-jointed to give as sleek and smooth an external appearance as possible. The cladding had suffered a

Fig. 8.12 Burleigh School, Hertfordshire, 1947, one of the schools designed by the Herfordshire County Council Architect's Department, with a design based around child-like proportions. A modular prefabricated construction method was developed to provide this series of small-scale buildings (photo Susan Macdonald).

number of problems over the years, and replacement of all the panels was regarded as inevitable during the 1995 investigations. The manner in which the buildings were erected meant that the replacement of individual panels was problematic, as the entire frame needed to be disassembled to remove one panel.

Analysis of the concrete problems revealed the following issues:

- the thinness of the panels had led to early carbonation of the concrete to the level of the reinforcement, causing reinforcement corrosion and resulting in cracking and spalling (Fig. 8.13);
- the corrosion of the steel fixings that hold the panels to the frame caused further spalling of the panels around the fixings, and in some cases panels had popped off the frame (Fig. 8.14);
- the thinness of the panels had resulted in some bowing/warping of the panels;
- these was some minor erosion to the exposed concrete surface.

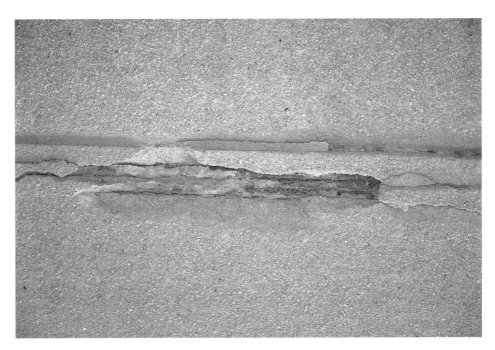

Fig. 8.13 Typical cracking of the concrete panels due to inadequate concrete cover (photo Susan Macdonald).

Fig. 8.14 Corrosion of the fixings at the corners and typical corrosion along the top and bottom of the precast panels. Note the junction with the window sill. Increasing the thickness of the panels creates a difficulty at the junction of the panels and the cover of the windows and door frames (photo Susan Macdonald).

The initial repair proposal was to replace the existing panels in exposed concrete precast panels 5 inches thick (127 mm) and chamfered at the edges. This was to provide the necessary cover to the reinforcement to prevent repeating the problem of early corrosion of the reinforcement. The chamfer was proposed as a means of accommodating the additional thickness of the panel within the lining of the building, and to accommodate the junction between the doors and windows given the extra panel thickness. It also made the panels easier to install under eaves. It was thought the chamfered edges would also be less likely to be damaged during their installation than the sharp straight edges of the original panels. However, the chamfered panels would have a significant visual effect on the buildings, giving a heavily rusticated appearance to what was intended to be a smooth streamlined elevation stripped of decoration.

Alternative solutions

After discussions with precast concrete manufacturers, it was established that the panels could be manufactured without the chamfered edges, and be butt-jointed, as was originally the case, without difficulty. By using stainless steel reinforcement, the thickness of the panels could be reduced to 3 inches (76 mm). Coffering the panels, with the excess thickness being accommodated in the lining/frame of the building, would provide added stiffness to prevent bowing of the panels. Improving the quality of the original concrete mix but replicating its colour and texture gave additional protection against early reinforcement corrosion. The use of a coating was discussed, but this would change the appearance of the building and therefore was not recommended.

Conclusions

The use of stainless steel reinforcement and high-performance concrete mixes increased the cost of the proposed repair. When considering the extended service life-cycle of the building which such an option provides, this may be a more cost-effective solution in the long term, while at the same time retaining the aesthetic significance of the building.

As the elevations of the building were deteriorating at different rates due to their different exposure levels, it was suggested that the repairs could be staged over a number of years, commencing with the most urgent elevations. This way the cost could be spread over a longer period—a further economic benefit.

University of Sussex: developing repair methods for the listed buildings

Introduction

The University of Sussex was designed by Sir Basil Spence during the 1960s. The complex of buildings, set within a park-like setting in the South Downs, uses a palette of red brick and board-marked concrete. The bold structures take on a solid and sometimes monumental form, and the materials are exuberantly expressed, providing texture and colour to the landscape. A number of the buildings at the University are now listed; Falmer House (1960–1962) as Grade I and eight others as Grade II.

The university has its own in-house team responsible for maintenance and on-going works to the university buildings. As listed building consent is required for works to most of the interior and exteriors of the listed buildings, English Heritage developed a close working relationship with the university in order to manage this process in the best possible way. The on-going programme of repair and maintenance sought to address the problems associated with the concrete. Since the buildings were listed, they have also needed to deal with the difficulties of reconciling the necessary concrete repair works with conservation requirements such as the need to minimise the impact of the repairs on the appearance of the buildings.

Concrete repair works

The concrete decay mechanisms in the buildings around the university can be broadly categorised as:

- reinforcement corrosion caused by poor initial construction, i.e. low cover to reinforcement, poor placement of reinforcement, inadequate site supervision;
- reinforcement corrosion in some areas due to chloride ingress as a result of the proximity to the sea;
- reinforcement corrosion due to the use of calcium chloride as an additive in some of the later buildings.

A major programme of concrete repair was initiated during the 1990s. The Meeting House (1963–1965), an exposed concrete building, was repaired prior to the buildings being listed in 1992. The repair used a traditional patch repair approach, i.e. cutting out areas of patent and latent damage, cleaning the rebars and patch repairing. An anti-carbonation coating was

Fig. 8.15 The Meeting House at the University of Sussex following its repair 2 years earlier in 1992. The building was coated with an anti-carbonation coating. Although this attempts to replicate the colour of the concrete, the finish gives a plastic shiny appearance and some of the texture of the board-marking has been lost.

then applied in a colour approximating to the existing concrete colour (Fig. 8.15). The coating provides some added protection against continuing deterioration of the concrete due to reinforcement corrosion, but has affected the building's appearance. This was the only concrete building in Spence's complex, and the exposed board-marked concrete provided a contrast with the red brick buildings that it sits between.

The Mathematics and Physical Sciences Buildings (1960–1962) were repaired in 1995. On the basis that there was no alternative solution, these buildings were repaired the same way as the Meeting House. However, experimentation with the coatings achieved a better match with the original concrete colour. The application of the coating requires fairly aggressive cleaning of the surface, which removes some of the surface texture of the board-marked finish. In addition, the application of a 'fairing coat' to fill any voids and provide a more continuous surface for the coating also smooths the surface, resulting in a flatter, less-textured and 'plastic' looking finish.

Falmer House was the next listed building to be considered and, given its Grade I listed status, attempts were made to develop a repair that did not obscure the rich expressive language of the brick and exposed concrete. It was decided not to repair the building all in one go, and that the first stage would be limited to one corner of the courtyard building.

Analyses of the areas of patent and latent damage on the quarter of the building to be repaired indicated that the concrete was in good condition and the areas requiring patch repair were minimal. It was also agreed that given that the carbonation on the front had not yet reached critical levels, the application of a coating was not immediately necessary.

Working with SIKA, a well-known manufacturer of concrete repair products that the university had involved fairly early on in the project, a palette of colour-matched polymer-modified repair mortars were developed. Although the repairs are still visible on careful inspection, they do meet the conservation aims appropriate for such an important building far more satisfactorily than the standard approach (Fig. 8.16). Refinement of the placement and finishing of the patches would further improve the repairs. As part of the repairs to the Alexandra Road Housing Estate in London, for example, latex moulds were made to replicate the board-marked finish. The university's intention was to develop a standard specification for the patch repairs that was agreed by English Heritage and which could act as a basis for future works.

This carefully crafted approach to concrete repair is unusual, and is outside the general 'culture' of concrete repair. However, the extra care required in colour matching and finishing the patches is warranted in some instances. This, of course, is standard practice in the repair of historic stonework, and the approach can inform concrete repair for listed buildings.

There are concerns about patch repairs changing colour over time or standing out when wet. This may be the case where the original aggregates and cement cannot be matched, but the repairs may also weather-in over time and become less noticeable. All buildings weather over time and establish their own patina.

More recently, the university initiated trial repairs using migratory corrosion inhibitors. Visually these had no effect on the appearance of the building in the short term. However, as their long-term performance is still largely unknown, their use was limited to those buildings that are unlisted.

Fig. 8.16 Falmer House after repair. The crack repair through the beam is not particularly successful. The repair could have been improved visually by improved finishing. The other patch repairs to this corner of the building have blended in well with the parent concrete.

Conclusions

The concrete repairs at the University of Sussex have evolved as the importance of the buildings has become better understood. Attempts have been made to initiate repairs that are both functional and preserve the buildings' special historic interest. As the concrete repair industry responds to the special demands which conservation requires, experience will grow and other approaches will undoubtedly be developed.

Cathodic protection of the National War Memorial Carillon Tower, Wellington, New Zealand

John Broomfield, W.R. Jacobs, W.L. Mandeno & S.M. Bruce

Introduction

New Zealand's National War Memorial is located in Wellington and consists of the Carillon, completed in 1932 (Fig. 8.17), and the Hall of

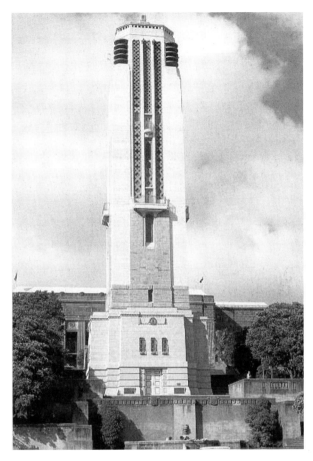

Fig. 8.17   The National War Memorial Carillon Tower, Wellington, New Zealand (photo Willie Mandeno).

Memories, which was consecrated in 1964. The memorial commemorates the New Zealanders who gave their lives in the South African War, in the First and Second World Wars, and in the wars in Korea, Malaysia and Vietnam; it also honours the New Zealand men and women who served in those wars.

The Carillon is a 50-m-high reinforced concrete structure which contains 74 bells that range in weight from 10 kg to 12.5 tonnes. Their total weight of 70.5 tonnes and its six-octave range makes it the third largest carillon in the world.

Despite repairs carried out during seismic strengthening in 1982, the Carillon was suffering from on-going concrete deterioration. In 1988, a condition assessment of the concrete structure was carried out with a view to developing appropriate repair options.

Condition assessment

A condition survey revealed that the principal form and area of deterioration was reinforcement corrosion in the belfries. The internal concrete surfaces are exposed to moisture and marine aerosol carried by the prevailing northerly and southerly winds through the belfry openings. A significant contributing factor was poorly compacted concrete and the limited protection it gave to the reinforcing steel. Testing indicated that carbonation depths of up to 70 mm were contributing to the corrosion. The concrete adjacent to the openings was also suffering from chloride contamination. Cracking and spalling was fortunately not widespread, and was largely confined to the concrete mullions and reveals of the openings. The outer surfaces of the concrete had a low moisture content due to the protection afforded by the stone cladding and plaster work.

Repair options

Two remedial options were developed following this initial assessment.

1. A conventional patch repair system followed by a protective coating.
2. A combination of the electrochemical repair techniques of realkalisation and electrochemical chloride extraction (ECE), also known as desalination.

Option 1 was the cheaper alternative, but it was predicted that further patch repair might be required within 10–20 years. The realkalisation process reinstates the alkalinity of the concrete which is lost due to carbonation, and the ECE extracts some of the chloride ions from the concrete and repassivates the steel. As these repair techniques reverse the causes of corrosion, it was predicted that a maintenance-free life of more than 25 years could be achieved.

Owing to the significance of the Carillon as a national monument and protected heritage structure, the client—the Department of Internal Affairs—chose the electrochemical repair option since it offered the longest maintenance-free life. A trial installation of realkalisation and ECE was then commissioned to assess the feasibility of these techniques to reinstate alkalinity and remove chloride ions in the Carillon belfries.

In practical terms, both the electrochemical processes involve the initial repair of spalled or cracked concrete, and the attachment of an anode system (usually a mixed metal oxide-coated titanium mesh in an electrolyte) to the surface. The appropriate voltage is then applied between the mesh and the reinforcing steel. Finally, the anode and electrolytic mixture are removed once the process has been completed. The extent and quality of repair required before these electrochemical techniques are applied is far less than a conventional patch repair because only the section of concrete that has spalled needs to be reinstated.

Trials

Realkalisation and ECE are available in New Zealand as the proprietary 'Norcure' process, that is owned by Fosroc International. As a licensee for this process, Consultech (a division of Construction Techniques Group Ltd.) was commissioned to design and install a trial installation to determine the operating parameters for a possible full-scale ECE and realkalisation treatment on the structure. Four separate areas in the upper belfry were treated.

The results showed that the trial installation was successful in its objective of significantly lowering chloride levels around the reinforcing bars and re-establishing an alkaline environment around the reinforcement steel in the areas treated. Additional information gained during the trial installation included:

- the fact that the chloride ion contamination was found to be more widespread than was originally estimated, based on additional testing carried out to monitor the ECE trials;
- the presence of an acrylic coating on surfaces from earlier repairs whcih lengthened the treatment process;
- the fact that a method of realkalisation using electrolyte-filled cassettes was more effective than the more usual spray-applied cellulose fibre owing to the dryness of the concrete and the logistics of spraying wet cellulose fibre in close proximity to the Carillon's bells.

The discovery that the chloride contamination was more extensive than was originally estimated meant that cathodic protection (CP) also became a viable remedial option.

Selection of the repair option

After obtaining cost estimates for the various electrochemical options, it was determined that the most economic repair method was patch repair followed by cathodic protection. To meet budgetary constraints, the CP system was specified to be applied only to those areas shown by a half-cell potential survey to be actively corroding.

Design of the repair system

Anode system

A number of anode systems were considered. Given the requirements, the exposure conditions and the condition of the concrete surface, it was decided to use probe anodes embedded in the concrete.

Trial installation of the CP system

The two trial areas were chosen as being representative of the areas to be cathodically protected. Half-cell reference electrodes were embedded below the lowest anode and between the top two anodes. The system was subject to half-cell potential surveys to determine the effectiveness of the treatment.

The trials determined that a further adjustment of the system was required to achieve the requisite polarisation of the steel. It was concluded that the level of protection required 100-mm-long 'Duranodes' at 100-mm spacing, powered at 1 mA per anode once the system was fully polarised, which could take several months. The mullions required additional protection, and here 150-mm-long 'Duranodes' at 600-mm spacing were selected, applying 3 mA per anode.

System installation

The impressed CP system installed utilises 570 discrete internal anodes. Mixed metal oxide-coated titanium rods, 3 mm in diameter, were inserted in carbon gel 600 mm apart. The anodes were generally 100 mm long, or sometimes 150 mm long, in the mullions, as described above. Twenty-five zones were established, each with its own embedded Ag/AgC reference electrode, to give permanent protection to approximately 170 m² of at-risk structural concrete. Holes 12 mm in diameter were drilled 600 mm apart, as required The holes were checked with a 'down hole cover meter' to ensure that they were not too close to the reinforcement. They were then filled with the carbonaceous back-fill and the anode rod was inserted. A titanium wire was connected between the anodes and wired into a junction box, and then connected back to the power supply. A remote monitoring system was installed to monitor the 25 zones.

Commissioning the system

The CP system was commissioned in June 1999. The system is remotely controlled, and monitored by the contractor who installed the system from their Auckland office. The commissioning results indicted that some zones were failing to achieve their targets. Minor adjustments were made to the system, including increasing the voltage level to the system from 10 to 12 V. The system is now showing a high level of agreement with the control criteria in EN 12696 (2000), the new European Standard on cathodic protection of steel in concrete.

Conclusions

Careful assessment of the concrete deterioration problems at the Carillon identified the causes of decay and the extent of the problem, and provided a risk assessment on its continued deterioration. The repair options initially selected aimed at reducing on-going deterioration and minimising invasive repairs to this important heritage structure. As the condition of the reinforcement corrosion and the extent of chloride contamination became evident during the trial repair phases, it was decided that an alternative approach could better address the deterioration problems. The selection of a cathodic protection system provides an on-going maintenance function by preventing continued corrosion of the reinforcement steel in the maritime environment.

Building pathology: concrete

The range of approaches illustrated in these case studies highlights the necessity of adopting a case-specific solution driven by the causes of the defect, damage or decay that has affected the structure. Other issues such as conservation requirements, occupancy issues, aesthetics, the availability of various repair options and, inevitably, the cost will also play a role. However, it is vital to consider whatever approach is adopted as part of a long-term maintenance strategy for the building.

The building pathology approach is summarised below.

1. Understand the building, i.e. how it is constructed, the materials of which it is constructed, how it has been used, and how it has evolved over time.
2. Understand the environment, and the effect of that environment on the physical condition of the building.
3. Identify the defects, damage and decay, and their causes.
4. Understand the repair options available.
5. Ensure that maintenance strategies are put in place that will safeguard the building in the future and minimise further deterioration.

This comprehensive approach to building care will inevitably assist in sustaining the building more effectively to allow its use to continue into the future.

References

Architectural Forum (1950) 'Chicago Apartment Developments: Mies van der Rohe's Promontory and Lake Shore Projects: Glass and Brick in a Concrete Frame,' *Architectural Forum* (January 1950): 69–77.

ASTM C856-95e1 (1995) *Standard Practice for Petrographic Analysis of Hardened Concrete.* American Society for Testing and Materials, West Conshohocken, PA.

ASTM C231-97e1 (1997) *Standard Test Method for Air Content of Freshly Mixed Concrete by the Pressure Method.* American Society for Testing and Materials, West Conshohocken, PA.

ASTM C114-00 (2000) *Standard Test Methods for Chemical Analysis of Hydraulic Cement.* American Society for Testing and Materials, West Conshohocken, PA.

ASTM C39/39M-01 (2001) *Test Method for Compressive Strength of Cylindrical Concrete Specimens.* American Socity for Testing and Materials, West Conshohocken, PA.

ASTM C143/C143M-00 (2001) *Standard Test Method for Slump of Hydraulic Cement Concrete.* American Society for Testing and Materials, West Conshohocken, PA.

EN 12696 (2000) *Cathodic protection of steel in concrete.* European Standard, BSI, London.

Further reading

Allen, J. (1994) The conservation of modern buildings. In: Mills, E. (ed) *Building Maintenance and Preservation: A Guide to Design and Management,* revised edition. Butterworth–Heineman, London.

Bronson, S. and Jester, T. (eds) (1997) Special issue: mending the modern. *APT Bulletin* **28**:4.

De Jonge, W. and Doolaar, A. (1997*) The Fair Face of Concrete: Conservation and Repair of Exposed Concrete.* Preservation Dossier No. 2, DOCOMOMO International, Eindhoven.

Leatherbarrow, D. and Mostafari, J. (1993) *On Weathering: The Life of Buildings in Time.* MIT Press, Cambridge.

Macdonald, S. (ed) (1996*) Modern Matters: Principles and Practice in Conserving Recent Architecture.* Donhead, Shaftesbury.

Macdonald, S. (ed) (2001) *Preserving Post-War Heritage: The Care and Conservation of Mid-Twentieth-Century Architecture.* Donhead, Shaftesbury.

Slaton, D. and Foulks, G. (2000) *Preserving the Recent Past, Vol. 2.* Historic Preservation Education Foundation, Washington, DC.

Slaton, D. and Shiffer, R.A. (eds) (1995) *Preserving the Recent Past.* Historic Preservation Education Foundation, Washington, DC.

Stratton, M. (ed) (1997) *Structure and Style: Conserving Twentieth-Century Buildings.* E & FN Spon, London.

Appendix A
Useful Addresses

American Concrete Institute
ACI International
PO Box 32431
Detroit, MI 48232-0431
USA
Tel: +1 248 848 3800
Fax: +1 248 848 3801
Website: http://www.aci-int.org
ACI International is a professional organisation with a membership in the fields of concrete design, construction, materials, education and certification. It publishes a number of periodicals, including *Concrete International Magazine*, *ACI Structural Journal* and *ACI Materials Journal*.

American Society for Testing and Materials (ASTM) Standards Coordination
100 Barr Harbor Drive
West Conshohocken
PA 19428-2959, USA
Tel: +1 610 832 9500
Fax: +1 610 832 9555
Website: http://www.astm.org

Association for Preservation Technology International (APTI)
4513 Lincoln Ave., Suite 213
Lisle, IL 60532-1290, USA
Tel: +1 1 630 968 6400
Fax (toll free): +1 1 888 723 4242
e-mail: information@apti.org
Website: http://www.apti.org
APTI is an interdisciplinary membership organisation dedicated to the practical application of the principles and techniques necessary for the maintenance, conservation and protection of historic buildings, sites and artifacts. It produces regular newsletters and journals, convenes conferences and other training courses, and has stimulated much discussion and published material on the subject of the conservation of historic concrete over the last 10 years.

British Cement Association (Centre for Concrete Information)
Century House
Telford Avenue
Crowthorne, RG45 6YS, UK
Tel: +44 1344 762676
Fax: +44 1344 761214
e-mail: library@bca.org.uk
Website: http://www.bca.org.uk/
The British Cement Association offers expert advice in response to enquiries on concrete and cement. It has a comprehensive library, including a historical section. The library also holds two databases on products and services, and on literature. Membership of the BCA includes free literature searches and free access to product and technical information, as well as access to the library.

British Precast Concrete Federation
60 Charles Street
Leicester LE1 1FB, UK
Tel: +44 116 253 6161
Fax: +44 116 251 4568

e-mail: info@britishprecast.org
Website: http://www.britishprecast.org.uk/
For information on manufacturers of precast units, including historical information.

British Standards Institution (BSI)
389 Chiswick High Road
London W4 4AL, UK
Tel: +44 20 8996 9000
Fax: +44 20 8996 7400
e-mail: info@BSI_global.com
Website: www.bsi-global.com
In operation since 1901, BSI is the national standards body in the UK, but also operates internationally. It provides standards for a wide range of industries and management systems, and is a testing house for products and materials. It publishes about 1700 standards a year and provides an online search engine, and BSI documents are available online.

Building Research Establishment (BRE)
Garston, Watford
Herts WD2 7JR, UK
Tel: +44 1923 664000
Fax: +44 1923 664010
e-mail: enquiries@bre.co.uk
Website: http://www.bre.co.uk
After 75 years in the public sector, BRE became a private company in March 1997 and is now owned by the Foundation for the Built Environment. The Foundation is a non-profit distributing body with around 150 members made up from representative firms, professional bodies and other organisations drawn from a wide spectrum of construction and building owners. It was created to ensure that BRE remains independent of commercial interests, and retains its reputation for objectivity and impartiality in research and consultancy.

BRE is the UK's leading centre of expertise on building and construction, with current capabilities covering many aspects of building performance—environmental,

structural, materials durability and fire safety. It has a long and distinguished history starting from the 1920s and has, over this period, developed a large knowledge base derived from independent high-quality research. Until recently, BRE's main role was to advise and carry out research for the UK Government. It also provides extensive research and development services directly to industry, and has an active publishing, conference and information arm.

Concrete Repair Association
Association House
235 Ash Road
Aldershot GU12 4DD, UK
Tel: +44 1252 321302
Fex: +44 1252 333901
e-mail: info@associationhouse.org.uk
Website: www.concreterepair.org.uk
The Concrete Repair Association promotes and develops the practice of concrete repair, advances education and technical training, and represents its members to professional bodies, authorities and specifiers. Membership comprises contracting companies and manufacturers/suppliers established in concrete repair.

The Concrete Society Ltd.
No 3 Eatongate
112 Windsor Road
Slough SL1 2JA, UK
Tel: +44 1344 466007
Fex: +44 1344 466008
e-mail: concsoc@concrete.org.uk
Website: http://www.concrete.org.uk
First established in 1966, The Concrete Society is an independent member-based organisation working to develop and promote excellence in concrete. The Society's membership includes leading professionals and trade personnel. It now offers a web-based information service to members and the public.

Corrosion Prevention Association (CPA)
Association House
235 Ash Road
Aldershot GU12 4DD, UK
Tel: +44 1252 321302
Fex: +44 1252 333901
e-mail: info@associationhouse.org.uk
Website: www.corrosionprevention.org.uk
This society aims to identify, quantify and communicate the principles and benefits of cathodic protection, to promote its use where it is the correct technical and commercial solution, and to maintain the highest professional standards of conduct and quality in all aspects of cathodic protection of reinforced concrete. It offers a list of consultants and contractors who are able to carry out cathodic protection.

Corrosion and Protection Centre
UMIST
PO Box 88
Manchester
M60 1QD, UK
Tel: +44 161-200-4848
Fax: +44 161-200-4865
e-mail: corrosion@umist.ac.uk
Website: http://www.cp.umist.ac.uk/CPC/
The Centre was established in 1972 at UMIST as an interdisciplinary centre that encourages specialists from different academic disciplines to cross their traditional boundaries to work in relevant areas of corrosion and protection. The Centre has three main areas of activity in corrosion and protection, namely teaching, research, and advisory and testing services for industry.

DOCOMOMO International
Delft University of Technology
Faculty of Architecture
Berlageweg 1
2628 CR Delft
The Netherlands
Tel: +1 31-15-2788755
Fax: +1 31-15-2788750

e-mail: docomomo@bk.tudelft.nl
Website:
http://www.bk.tudelft.nl/docomomo/
DOCOMOMO International (the international working party for the documentation and conservation of building sites and neighbourhoods of the Modern Movement) aims to promote recognition of the Modern Movement, and is actively involved in the identification and promotion of distinctive Modern Movement architecture nationally and internationally, acting as advisors to ICOMOS International on the World Heritage listing of modern buildings. Through their specialist committees (including a technical committee), they are actively involved in the development of appropriate conservation measures, including protection, technical development and dissemination of knowledge. They also attempt to identify and attract funding for documentation and conservation to ensure continued exploration and development of Modern Movement architecture. There are currently more than 30 member countries. A quarterly journal and biennial international conferences are some of the benefits to members. There is also a database of current technical research into the conservation of Modern Movement buildings.

ICOMOS
49-51 rue de la Fédération
75015 Paris, France
Tel: +1 33 1 45676770
Fax: +1 33 1 45660622
e-mail: secretariat@icomos.org
Website:
http://www.international.icomos.org
ICOMOS (International Council on Monuments and Sites) is a non-governmental professional organisation formed in 1965. It is closely linked to UNESCO, with members in over 60 countries (including the UK) and headquarters in France. Members in these countries are formed into National Commit-

tees and have the right to participate in the General Assemblies of ICOMOS, which are held triennially. ICOMOS is primarily concerned with the philosophy, terminology, methodology and techniques of cultural heritage conservation.

English Heritage
PO Box 569
Swindon, SN2 2YP
Tel: +44 870 333 1181
Fax: +44 1793 414 926
e-mail: customers@english-heritage.org.uk
Website: http://www.english-heritage.org.uk/
English Heritage is an independent body sponsored by the Department of Culture, Media and Sport. Its aim is to protect England's architectural and archaeological heritage for the benefit of present and future generations. It is the country's principal adviser on the historic environment, and is responsible for advising on the listing and scheduling of buildings and monuments to ensure legal protection. **Historic Scotland** (www.historic-scotland.gov.uk), **Cadw** (www.cadw.wales.gov.uk) and **Northern Ireland Environment and Heritage Service** are the equivalent organisations for other parts of the UK.

The Institution of Civil Engineers (ICE)
Great George Street
London SW1, UK
Tel: +44 20 7222 7722
Fax: +44 20 7222 7500
Website: www.ice.org.uk
The ICE is a UK-based international professional organisation of civil engineers. It provides a range of educational and publishing services, including an online journal, *Structural Concrete*, that seeks to cover all aspects of the design, construction, performance in service and demolition of concrete structures.

The Institute of Corrosion (Icorr)
4 Leck House
Lake Street
Leighton Buzzard LU7 8TQ, UK
Tel: +44 1525 851771
Fax: +44 1525 376690

International Concrete Repair Institute (ICRI)
1323 Shepard Dr., Suite D
Sterling, VA 20164-4428
USA
Tel: +1 703 450 0116
Fax: +1 703 450 0119
e-mail: concrepair@aol.com
Website: www.icri.org

The Institution of Structural Engineers
11 Upper Belgrave Street
London SW1X 8BH, UK
Tel: +44 20 7235 4535
Fax: +44 20 7235 4294
e-mail: istructe.org.uk
Website: http://www.istructe.org.uk
The Institution of Structural Engineers is an authorised and nominated body of the Engineering Council and a founder member of the Construction Industry Council. The Institute publishes a yearbook of members and a directory of firms for the UK. Courses, including concrete repair, are offered on a regular basis.

National Association of Corrosion Engineers
(NACE International)
PO Box 218340
Houston, TX 77218-8340, USA
Tel: +1 281 228 6200
Fax: +1 281 228 6300
Website: http://www.nace.org

Prestressed Concrete Association
60 Charles Street
Leicester LE1 1FB, UK
Tel: +44 116 253 6161
Fax: +44 116 251 4568
e-mail: info@britishprecast.org

RILEM
Ecole Normal Supérieure
Pavillion des Jardins
61 Avenue du Pdt. Wilson
F-94235 Cachan Cedex
France
Tel: +33 1 47 40 23 97
Fax: +33 1 47 40 01 13
e-mail: pascale.callec@rilem.ens-cachan.fr
Website: www.rilem.org
A non-profit technical organisation aimed at progressing knowledge in construction science and practice, and promoting information exchange between experts world-wide.

Scandinavian Society for Corrosion Engineering
PO Box 13034, S-25013, Helsingborg, Sweden
Tel: +46 42 16 22 78
Fax: +46 42 16 29 92

Society for the Protection of Ancient Buildings
37 Spital Square
London E1 6DY, UK
Tel: +44 20 7377 1644
Fax: +44 20 7247 5296
e-mail: info@spab.org.uk
Website: http://www.spab.org.uk
The Society for the Protection of Ancient Buildings was founded by William Morris in 1877 to counteract the highly destructive 'restoration' of medieval buildings being practised by many Victorian architects. The SPAB lobby to save old buildings from decay, demolition and damage, provide advice, host various educational programmes and produce a range of publications on the repair of historic buildings. Their approach is based on the principles encompassed in their manifesto.

Transportation Research Board
2101 Constitution Avenue NW
Washington, DC 20418, USA
Tel: +1 202 334 2379
Fax: +1 202 334 2006
Website: www.nas.edu/trb/
The Transportation Research Board (TRB) is a unit of the National Research Council (US) non-profit institution that is the principal operating agency of the National Academy of Sciences and the National Academy of Engineering. The Board's mission is to promote innovation and progress in transportation by stimulating and conducting research, facilitating the dissemination of information, and encouraging the implementation of research results.

Twentieth-Century Society
70 Cowcross Street,
London EC1H 6PP, UK
Tel: +44 20 7250 3857
Fax: +44 20 7251 8985
e-mail:
administrator@c20society.demon.co.uk
Website: www.c20society.demon.co.uk
Originally called the Thirties Society, the Twentieth-Century Society was founded with the aim of promoting and protecting British architecture and design after 1914. It conducts an annual lecture programme, visits, a newsletter and journals on twentieth-century design and architecture. Through casework, the Society assists statutory bodies in protection of twentieth-century architecture.

Glossary

Alkali–aggregate reaction (abbreviated as AAR) Reaction between the aggregates and the alkaline cement paste, leading to the development of an expansive crystalline gel which is sufficiently strong to cause cracking of the aggregate and of the concrete matrix. (Sometimes also called alkali–silica reaction, ASR.)

Anode The positive pole of an electric circuit in a cathodic protection system: a sacrificial material introduced to act as the site of corrosion to inhibit corrosion of the structure itself.

Bending moment The equal and opposite compressive and tensile forces acting inside a structural element to cause flexure when subjected to external forces, such as loading on a beam, or other actions, such as differential settlement along a strip footing.

Binder (in coatings) Polymeric complexes that provide integrity to the dried film and bond it to the surface to which it is applied. Binders are often referred to as resins or polymers.

Binder (in concrete) The materials that comprise the cementing agents in concrete, mortars and renders. Historically, 'natural' cements, such as the volcanic ash known as pozzolan, or lime or gypsum products, were used as binders. In concrete made during the last century or so, Portland cement (*q.v.*) has almost universally been the binder, although other artificial cements are increasingly being used which employ recycled industrial waste material such as ground granulated blast-furnace slag. Cement is mixed with water and added to aggregates (the **filler**, *q.v.*) to make concrete.

Building pathology Identification, investigation and diagnosis of defects in existing buildings, the prognosis of defects and recommendations for the most appropriate course of action having regard to the building, its future and the resources available, and the design, specification, implementation and supervision of appropriate programmes of remedial work, with monitoring and evaluation in terms of functional, technical and economic performance in use (Watt, 1999, given in Chapter 1).

Carbonation Loss of alkalinity in the concrete as a result of calcium hydroxide depletion (brought about by the presence of atmospheric carbon dioxide, which with moisture forms carbonic acid). The reaction of calcium hydroxide with the acid results in the formation of calcium carbonate, which neutralises the alkalis in the pores and results in the loss of the passivating oxide layer around the embedded steel in the carbonated zone. The **carbonation front** is the interface between the uncarbonated (virgin) concrete and the carbonated concrete as carbonation progresses inwards towards the steel. Carbonation is progressive, but occurs at a rate which reduces with time. It progresses faster in zones of local defects, such as cracking and poor compaction, than in the general body of (competent) concrete.

Cathode The negative pole of an electric circuit in a cathodic protection system: the metal is protected against corrosion by the presence of a sacrificial anode.

Cement The binding material that is one the components of concrete and is most commonly **Portland cement** (*q.v.*).

Chlorides As occurring in calcium chloride (used as a cement-setting accelerator in the past) and sodium chloride (in sea-water, wind-blown sea spray and road de-icing salt), chlorides combine with water to form an aggressive agent leading to accelerated corrosion of reinforcement.

Corrosion 'Rusting' or the formation of iron oxides and other compounds by electrolytic action when steel is exposed to water and oxygen. Corrosion is aggravated by other aggressive agents such as acids and chlorides (*q.v.*). Rust occupies a larger volume than the original iron, and consequently can cause cracking and spalling (*q.v.*) in the surrounding concrete.

Cost–benefit analysis An assessment of the desirability of projects, where the indirect effects on third parties outside those affecting the decision-making process are taken into account (Watt, 1999).

Cover The concrete between the reinforcement and the adjacent face of the element. It provides protection for the steel from corrosion (*q.v.*). The required thickness of cover and the quality of the concrete mix used are influenced by the severity of exposure, and must be chosen correctly to ensure durability (*q.v.*).

Creep The long-term shortening or deflection of the concrete as the strain increases under sustained stress, which usually has to be allowed for in the structural design of reinforced concrete.

Delamination Separation of layers of concrete from the main body of the material.

Design service life The anticipated time in service until deterioration processes have degraded the ability of the structure to meet the specified performance requirements to the point where a pre-defined unacceptable state is reached.

Durability Ability of a building and its parts to perform its required functions over a period of time and under the influence of internal and external agencies or mechanisms of deterioration and decay (Watt, 1999).

Economic service life The time in service until replacement is economically more advantageous than continued maintenance in service.

Electrochemical reaction Reaction involving ions in solution (Watt, 1999).

Filler The aggregates, typically coarse aggregate (crushed stone, gravel, etc.) and fine aggregate (commonly sand) which, when mixed with the binder (*q.v.*) and water, result in concrete.

Freeze–thaw degradation Occurs when moisture ingresses into the concrete, and in cold conditions freezes at and/or just below the surface (usually the first few millimetres beneath the surface). Ice occupies a greater volume than water, so with numerous freeze–thaw cycles the surface layer of the concrete becomes detached, resulting in spalling (*q.v.*). Over time, freeze–thaw conditions can continue to damage the concrete more deeply as progressive layers are detached.

Functional service life The time in service until the structure becomes functionally obsolete due to changes in performance or other requirements.

Galvanic action Occurs when two dissimilar metals are placed together in solution. The most active metal will become an anode and will be corroded as a current passes between the two metals. This action is used to stop corrosion by galvanising (coating steel with zinc), and in galvanic cathodic protection.

Impressed current cathodic protection A method of cathodic protection that uses a power supply and an inert (or controlled consumption) anode to protect a metallic object or element by making it the cathode.

In situ concrete Concrete cast in its intended location; cf. precast concrete (*q.v.*).

Intelligent design Means considering each structure on the basis of the specific performance requirements and the conditions of its use, rather than rigidly adhering to the generalised prescriptive rules in a code of practice.

Latent damage Non-visible damage that is impairing, or will, impair the functionality of the structure and will eventually require some form of remedial action.

Life cycle Time-interval that starts with the initiation of a concept and ends with the disposal of the asset (Watt, 1999).

Life-cycle costs Total costs of ownership of an item, taking into account all the costs of acquisition, personnel training, operation, maintenance, modification and disposal, for the purpose of making decisions on new or changed requirements, and as a control mechanism in service for existing and future items.

Maintenance A (usually) periodic activity or process intended either to prevent or correct the effects of minor deterioration or mechanical wear of the structure or its components in order to ensure their future functionality at the level anticipated by the designer. **Preventive maintenance** is maintenance carried out at predetermined intervals, or corresponding to prescribed criteria, and is intended to reduce the probability of failure or performance degradation of an item (Watt, 1999). **Emergency maintenance** is work carried out after some form of failure has occurred.

Mass concrete A term generally synonymous with unreinforced concrete (*q.v.*), but also applied to massive concrete elements and structures such as gravity dams, which may well have some reinforcement.

Nominal design life The service life intended by the designer, being the minimum period during which the structure, or a specified part of it, is intended to perform its design functions by meeting or exceeding the specified performance requirements, and subject only to routine servicing and maintenance.

Oxidation The process of removing electrons from an atom or ion. The process:

$$Fe \rightarrow Fe^{2+} + 2e^-$$

$$Fe^{2+} \rightarrow Fe^{3+} + e^-$$

is the oxidation of iron to its ferrous (Fe^2) and ferric (Fe^{3+}) oxidation states. Oxidation is done by an oxidising agent, of which oxygen is only one of many.

Passivation The process by which steel in concrete is protected from corrosion by the formation of a passive layer due to the highly alkaline environment created by the pore water. The passive layer is a thin, dense layer of iron oxides and hydroxides, with some mineral content, that is initially formed as bare steel is exposed to oxygen and water, but then protects the steel from further corrosion as it is too dense to allow the water and oxygen to reach the steel and continue the oxidation process.

Patent damage Visible damage in reinforced concrete decay: damage such as cracking, spalling, etc.

pH Logarithmic scale for expressing the acidity or alkalinity of a solution based on the concentration of hydrogen ions: a neutral solution has a pH of 7, whilst a pH below 7 indicates an acid solution and a pH above 7 indicates an alkaline solution (Watt, 1999). Concrete has a pH of 12–13. Steel corrodes at pH 10–11.

Pore (water) Concrete contains microscopic pores. These contain alkaline oxides and hydroxides of sodium, potassium and calcium. Water will move in and out of the concrete, and saturating, part filling and drying out the pores according to the external environment. The alkaline pore water sustains the passive layer if it is not attacked by carbonation or chlorides.

Portland cement Patented by Aspdin in 1824 and named after its resemblance to Portland stone. It is an artificial or manufactured material, although it is made from limestone or chalk, together with clay or shale. These ingredients contain alumina, silica, lime, iron oxide and magnesia, and are ground to a fine powder, burnt in a kiln, and then re-ground to a very fine powder which sets hard when mixed with water.

Post-tensioned concrete Prestressed concrete made by casting-in conduits or sheaths for prestressing steel that is tensioned and secured by anchorages once the concrete has cured. Profiling the conduits or sheaths produces a more efficient section, as noted below for pretensioned concrete. The conduits or sheaths are usually then grouted-up to provide a bond between the steel and the concrete and increase durability.

Precast concrete Reinforced concrete cast in moulds as units or elements, elsewhere than in their final intended location, before being placed into position.

Prestressed concrete Concrete that has had compressive stress applied to it by tensioned steel before it is put into service to carry loads. The prestressing steel may take the form of rods, wires, cables or bars. Prestressing increases the strength of the element and can eliminate cracking in service.

Pretensioned concrete Prestressed concrete made by tensioning the prestressing steel before the concrete is poured. This typically requires temporary anchorages to hold the ends of the steel, and stout moulds to resist the resulting compression forces exerted on them. Once the concrete has set, the anchorages are freed and the prestressing force is transferred as compression in the concrete. Pretensioning generally employs straight runs of steel, although sometimes it is profiled, following the pattern of the bending moment (*q.v.*) to give a more efficient use of the material.

Reinforced concrete Concrete reinforced with metal rods, straps, wires or mesh that provides a composite material which is strong in tension and compression. Nowadays, the reinforcement is most commonly mild or high-tensile steel, but iron, annealed wire, and galvanised and stainless steel have all been used in various ways as reinforcement. In the future, non-metallic high-strength composites used as reinforcement may reduce or eliminate concerns over corrosion and durability.

Repair action Taken to reinstate to an acceptable level the current functionality of a structure or its components that are either defective, degraded or damaged in some way.

Residual-service life The time between the moment of consideration and the end of service life when a predefined unacceptable state is reached.

Sacrificial anode cathodic protection A system of cathodic protection that uses a more easily corroded metal such as zinc, aluminium or magnesium to protect a steel object from corrosion. No power supply is required, but the anode is consumed.

Shrinkage Contraction of the cement paste as it hardens, due to loss of moisture and changes to the paste's internal structure. Some shrinkage is non-reversible due to these changes, while reversible shrinkage occurs as the concrete becomes wet in service and then dries again. Some materials that might otherwise be used as aggregates should be avoided if they are found to be 'shrinkable', as this property may damage the concrete in service.

Spalling Detachment of lens-like pieces of surface concrete, usually due to reinforcement corrosion and the production of expansive rust products that put the concrete locally into tension, resulting in cracking and then spalling.

Technical service life   The time in service when a predefined unacceptable state is reached.

Unreinforced concrete   Concrete that does not contain reinforcement.

Whole life-cycle cost   Generic term for the costs associated with owning and operating a facility from inception to demolition, including both initial capital costs ands running costs (Watt, 1999).

Index